Audio Production Basics with Ableton Live

To access online media visit:
www.halleonard.com/mylibrary
Enter code:

3332-0591-0336-8710

Audio Production Basics with Ableton Live

Eric Kuehnl

with contributions by
Frank D. Cook

NextPoint Training, Inc.
Ableton Live 10

Rowman & Littlefield

Lanham • Boulder • New York • London

Published by Rowman & Littlefield
An imprint of The Rowman & Littlefield Publishing Group, Inc.
4501 Forbes Boulevard, Suite 200, Lanham, Maryland 20706
www.rowman.com

6 Tinworth Street, London SE11 5AL, United Kingdom

Copyright © 2020 by NextPoint Training, Inc.

Book design by NextPoint Training, Inc.

All rights reserved. No part of this book may be reproduced in any form or by any electronic or mechanical means, including information storage and retrieval systems, without written permission from the publisher, except by a reviewer who may quote passages in a review.

Ableton Live and all related product names and logos are registered trademarks of Ableton AG in Germany and/or other countries. All other trademarks are the property of their respective owners.

The media provided with this book and accompanying course material is to be used only to complete the exercises contained herein or for other valid educational purposes. No rights are granted to use the media in any commercial or non-commercial production.

All product features and specifications are subject to change without notice.

Library of Congress Cataloging-in-Publication Data available
ISBN 978-1-5381-3756-7 (paperback)
ISBN 978-1-5381-3757-4 (eBook)

∞™ The paper used in this publication meets the minimum requirements of American National Standard for Information Sciences—Permanence of Paper for Printed Library Materials, ANSI/NISO Z39.48-1992.

Contents

Acknowledgments xiii

Introduction. Welcome to the World of Audio xv

Chapter 1. Computer Concepts 1
    Selecting a Computer 2
        Mac Versus Windows Considerations 3
        The Importance of RAM 3
        Processing Power 5
        Storage Options 8
        Onboard Sound Options (Audio In and Out) 10
        Other Options to Consider 10
    Working with Your Computer 11
        File Management 11
        Launching Applications 17
        Saving Files 18
    Working with an Application 19
        Menus 19
        Keyboard Shortcuts 21
    Review/Discussion Questions 23

Exercise 1. Exploring Audio on the Computer 25

Chapter 2. DAW Concepts 31
    Functions of a DAW 32
        What Can a DAW Do? 32
        Common DAWs 32
    Plug-In Formats 38
        What Is a Plug-In? 38
        Common Plug-In Formats 39
    Ableton Live Configurations 39
        Software Editions 39
        Audio Interface Options 41
    Registering Your Software 42

Software Installation and Operation 43
    Included Devices and Packs 44
    Installing Additional Packs 45
Launching Ableton Live for the First Time 47
    Launching Ableton Live 47
    Authorizing Live 47
    Accessing Connected Audio Devices 51
    Optimizing Ableton Live Performance 52
Important Concepts in Ableton Live 54
    Set Files Versus Audio Files 54
    Audio Versus MIDI 54
    Controlling Playback in Live 55
Review/Discussion Questions 57

Exercise 2. Setting Up a Multi-Track Project 59

Chapter 3. Audio Recording Concepts 65
The Basics of Audio 66
    Frequency 66
    Amplitude 67
Microphones 67
    Traditional Microphones 67
    USB Microphones 68
    Other Considerations 69
    Basic Miking Techniques 70
Multi-Tracking and Signal Flow 72
    What Is Multi-Track Recording? 73
    Recording Signal Flow 74
Moving Audio from Analog to Digital 74
    Analog Versus Digital Audio 74
    The Analog-to-Digital Conversion Process 75
The Audio Interface 76
    Audio Interface Considerations 76
    Working Without an Audio Interface 79
Review/Discussion Questions 80

Exercise 3. Selecting Your Audio Production Gear 82

Chapter 4. MIDI Recording Concepts 87
A Brief History of MIDI 88
Digital Control 88
The Birth of MIDI 89
The MIDI Protocol 90
MIDI Notes 90
Program Changes 91
Controller Messages 91
MIDI Controllers 92
Digital Pianos and Synthesizers 92
Keyboard Controllers 94
Drumpad Controllers 95
Grid Controllers 95
Alternate Controllers 96
What to Look for in a MIDI Controller 96
Purchasing a Keyboard Controller 97
Setup and Signal Flow 98
Plug-and-Play Setup 98
MIDI Cables and Jacks 98
Using a MIDI interface 99
Considerations for Using Multiple MIDI Devices 100
MIDI Versus Audio 100
What Is MIDI? 101
Monitoring with Onboard Sound Versus Virtual Instruments 101
Tracking with Virtual Instruments 102
Creating Tracks for Virtual Instruments 103
Summary 104
Review/Discussion Questions 105

Exercise 4. Selecting Your MIDI Production Gear 107

Chapter 5. Ableton Live Concepts, Part 1 111
Ableton Live Views and Sections 112
Control Bar 112
Arrangement View 113
Session View 114
Browser View 115
Detail View 116
Showing and Hiding Views and Sections 118

  Overview (Arrangement View only) 119
  Rulers and Scrub Area (Arrangement View only) 119
 Working with Tracks 120
  Adding Tracks 120
  Clips Versus Files 121
 Basic Navigation 121
  Playback in Arrangement View 121
  Using Follow Mode for Scrolling 124
 Zooming and Scrolling in the Arrangement View 125
  Zooming and Scrolling with the Beat-Time Ruler 125
  Zooming and Scrolling with the Arrangement Overview 125
  Zooming and Scrolling with Keyboard Shortcuts 126
  Optimizing Arrangement Height and Width 126
 Review/Discussion Questions 128

Exercise 5. Configuring and Working on a Project 130

Chapter 6. Ableton Live Concepts, Part 2 137
 Setting Up for Recording 138
 Recording Audio in Arrangement View 139
  Designating a Punch-In Point 139
  Record-Enabling Tracks 139
  Monitoring Record-Enabled Audio Tracks 140
  Initiating a Record Take 141
  Stopping a Record Take 141
 Recording MIDI in Arrangement View 142
  Monitoring a MIDI Controller 142
  Using MIDI Arrangement Overdub 143
 Importing Audio and MIDI 144
  Supported Audio Files 144
  Importing from the Desktop 145
  Importing from the Browser 145
  File Drag and Drop Locations 147
 Using the Edit Grid 147
  Edit Grid Modes 147
  Setting the Edit Grid Mode and Value 148
 Clip Editing in Arrangement View 149
  Basic Editing Techniques 149
  Advanced Editing Techniques 152

    MIDI Editing Techniques 155
        Editing MIDI Clips in the Arrangement View 155
        Using the MIDI Note Editor 155
        Navigating the MIDI Note Editor 159
    Review/Discussion Questions 162

Exercise 6. Importing and Editing Clips 164

Chapter 7. Session View 171

    The Session View 172
        Session View Clip Sections 172
        Session View Mixing Sections 175
    Adding Clips to Tracks in Session View 178
        Import to Tracks in Session View 178
        Import Directly to a New Audio Track in Session View 178
        Batch Importing Audio to Tracks in Session View 178
        Batch Import Directly to New Audio Tracks 179
    Playback in Session View 180
        Starting and Stopping Playback 180
        Controlling Playback with Scenes 181
        Enabling/Disabling Clip Looping 182
    Recording Audio in Session View 183
        Session View Recording Quantization 184
        Recording on Scene Launch 185
    Recording from Session View into an Arrangement 185
        Using Arrangement Record 185
        Playing Back an Arrangement Recording 185
    Review/Discussion Questions 187

Exercise 7. Working in the Session View 189

Chapter 8. Mixing Concepts 197

    Basic Mixing 198
        Setting Levels 198
        Panning 202
    Processing Options and Techniques 205
        Gain-Based Processing 205
        Time-Based Processing and Effects 206
        Inserted Devices Versus Sends 206
        Processing with External Gear 206

Mixing in the Box 207
    Advantages of In-the-Box Mixing 207
    Getting the Most Out of an In-the-Box Mix 208
Review/Discussion Questions 209

Exercise 8. Creating a Basic Mix 211

Chapter 9. Signal Processing 219
Processor Basics 220
    Viewing Devices on Tracks 221
    Inserting and Removing Devices 221
    Moving and Duplicating Devices 222
    Displaying Plug-In Windows 223
Common Device Controls 223
    Device Activator Button 224
    Hot-Swap Presets Button 224
    Save Preset Button 225
Adjusting Device Parameters 225
    Adjusting Device Parameters with the Mouse 225
    Adjusting Device Parameters with the Computer Keyboard 226
EQ Processing 226
    Types of EQ 226
    Basic EQ Parameters 227
    EQ Devices in Ableton Live 227
    Strategies for Using EQ 229
Dynamics Processing 230
    Types of Dynamics Processors 230
    Basic Dynamics Parameters 231
    Dynamics Devices in Ableton Live 231
    Strategies for Using Compression 232
Reverb and Delay Effects 233
    What Is Reverb? 233
    Reverb in Ableton Live 233
    Applications for Reverb Processors 234
    What Is Delay? 234
    Delay in Ableton Live 235
    Applications for Delay Processors 235

  Wet Versus Dry Signals 236
    Using Time-Based Effects Directly on Tracks 236
    Using Send-and-Return Configurations 237
  Review/Discussion Questions 238

Exercise 9. Optimizing Tracks with Signal Processing 240

Chapter 10. Finishing a Project 247
  Recalling a Saved Mix 248
  Automation 248
    Arrangement Automation Workflow 249
    Recording Arrangement Automation 249
    Viewing Arrangement Automation 250
    Drawing Automation Envelopes 251
    Editing Automation Breakpoints 253
    Deleting Automation 254
  Creating a Mixdown 255
    Adding Processing on the Master Track 255
    Considerations for Bouncing Audio 256
    Exporting Audio 256
  Review/Discussion Questions 261

Exercise 10. Preparing the Final Mix 263

Appendix A. Locating Missing Files in Ableton Live 269
  On-Screen Information about Missing Files 269
  Replacing via Drag-and-Drop 271
  Replacing via Automatic Search 271
  Dealing with Multiple Candidates 273

Index 277

About the Author 285

Acknowledgments

Eric would like to thank his wife, Amanda, and children, Thom and Freddie, for their support and understanding.

Special thanks go to the following individuals who have provided support, feedback, technical information, editorial input, and other contributions for this version of the book: John Cerullo, Andy Cook, Frank D. Cook, Matt Donner, Mark Garvey, Greg Gordon, and Bruce Tambling. Extra special thanks to the bands who provided media files used in the exercises herein: The Pinder Brothers and Fotograf.

Introduction

Welcome to the World of Audio

Sounds are all around us. They make our world interesting, informative, and engaging. It's a natural human desire to want to capture these auditory experiences—the sounds we like, the sounds we want to share, and the sounds we create. That is truly what audio production with Ableton Live is all about: creating, capturing, and sharing sound.

Ableton Live software provides a powerful springboard for exploring the fundamentals of audio production and music creation. The robust toolset included with this software package will inspire your creativity and help you produce great-sounding audio compositions. This book has been written for readers using any edition of Ableton Live software, from the entry-level Live Intro (or Live Lite), to the professional Live Standard, or the fully loaded Live Suite. Everything discussed in these pages applies equally to all editions of the software, unless otherwise noted. The included exercises include options that are compatible with the most basic Ableton Live Lite setup. No additional hardware or software is required.

With this book, you're taking the first step toward discovering the unique capabilities of Ableton Live software and unlocking your own audio creativity. Whether you've selected this book to use for self-study or have picked it up as the required text for instructor-led classroom training, you will find that it covers the principles of audio production from the ground up. It also gives you everything you need to know to fully understand the role of Ableton Live software in today's landscape of digital audio workstations (DAWs).

Getting Started in Audio Production

This book teaches the basics of recording, editing, mixing, and processing audio and MIDI using Ableton Live software. It also provides plenty of power tips to take you beyond the basics and unleash the true power of using Live as a creative tool. Additionally, the principles you learn in this book will apply equally to other commercial products used to create, record, edit, and process digital audio. This means you'll have a solid foundation for further growth, whether that involves upgrading your Ableton Live software, switching to another digital audio workstation, or working across multiple platforms in a full-fledged commercial studio.

Who Should Read This Book

Although this book supports entry-level editions of Ableton Live (Intro and Lite), the concepts apply equally to the full-featured Ableton Live Standard edition and the expansive Ableton Live Suite option. As such, this book serves as a resource for users of any edition of Live software. Learners who are running Ableton Live

Intro, Standard, or Suite can use this book and complete the included exercises as written. Those running Ableton Live Lite edition can use the alternate material provided to complete the same processes using a smaller compatible Set file.

To ensure that the topics covered in this book apply to all editions of Ableton Live software, we have included additional details on the differences between platforms where they apply. This book is designed to help new and inexperienced users get started with little or no background knowledge. However, the scope is not limited to novice users who are experimenting with DAWs for the first time.

About This Book

This book is designed for use in a formal course of study. The text and the associated instructor-led course were developed by NextPoint Training, Inc. (NPT) as part of our Digital Media Production program and certification offerings. While this coursebook can be completed through self-study, we also recommend the hands-on experience available through an instructor-led class with an NPT Certification Partner.

For more information on the classes offered through the NextPoint Training Digital Media Production program, please visit https://nxpt.us/DigitalMedia. For information on how your school can become an NPT Certification Partner, please visit https://nxpt.us/CertPartner.

Requirements and Prerequisites

This course does not require any specific background knowledge of computer systems, recording technology, or digital audio workstations. However, before starting to work with Ableton Live software, it definitely helps to have at least a passing familiarity with computer concepts and recording gear. If you consider yourself a novice in these areas, pay special attention to the first four chapters of this book.

To try out the concepts and complete the exercises in this book, you will need to use a compatible computer and install any current edition of Ableton Live software. Details on computer requirements are provided in Chapter 1, and installation details for Ableton Live are provided in Chapter 2.

If you will be using other connected audio or MIDI hardware, you may need to install device drivers for those components. Consult the documentation that came with your hardware or search the manufacturer's website for details.

Media Files

This book includes exercises at the end of each chapter that make use of various media files. The media files can be accessed by visiting www.halleonard.com/mylibrary and entering your access code, as printed on the opening page of this book. Instructions for downloading the media files are provided in Exercise 1.

Media files for the exercises in this book have been provided courtesy of The Pinder Brothers (Exercises 1, 2, and 5) and Fotograf (Exercises 6 through 10). Additional exercise media and remix files have been provided by Eric Kuehnl.

Course Organization and Sequence

This course has been designed to teach you how to get the most out of your work with Ableton Live software. The material is organized into 10 chapters, as follows:

- **Chapter 1. Computer Concepts**—What you need from a computer
- **Chapter 2. DAW Concepts**—What you need from your DAW
- **Chapter 3. Audio Recording Concepts**—What you need to record audio
- **Chapter 4. MIDI Recording Concepts**—What you need to record MIDI
- **Chapter 5. Ableton Live Concepts, Part 1**—What you need to know to get started with Ableton Live
- **Chapter 6. Ableton Live Concepts, Part 2**—What you need to know to work with Ableton Live
- **Chapter 7. Session View**—How to Make Use of the Ableton Live Session View
- **Chapter 8. Mixing Concepts**—What you need to know to mix a project
- **Chapter 9. Signal Processing**—What you can do to optimize your audio
- **Chapter 10. Finishing a Project**—What you need to do to create a stereo mixdown

Users who have experience with computers and other DAWs may wish to skim the first four chapters, focusing mostly on the details that are specific to Live software.

Conventions and Symbols Used in This Book

Following are some of the conventions and symbols we've used in this book. We try to use familiar conventions and symbols whose meanings are self-evident.

Keyboard Shortcuts and Modifiers

Menu choices and keyboard commands are typically capitalized and written in bold text. Hierarchy is shown using the greater than symbol (>), keystroke combinations use the plus sign (+), and mouse-click operations use hyphenated strings, where needed. Brackets ([ ]) indicate key presses on the numeric keypad.

Convention	Action
File > Save Session	Choose Save Session from the File menu.
Ctrl+N	Hold down the Ctrl key and press the N key.
Command-click (Mac)	Hold down the Command key and click the mouse button.
Right-click	Click with the right mouse button.
Press [1]	Press 1 on the numeric keypad.

Icons

The following icons are used in this book to call attention to tips, shortcuts, listening suggestions, warnings, and reference sources.

 Tips provide helpful hints and suggestions, background information, or details on related operations or concepts.

 Shortcuts provide useful keyboard, mouse, or modifier-based shortcuts that can help you work more efficiently.

 Power Tips provide shortcuts and tips for power users that can dramatically speed up your work but that go beyond the scope of the current discussion.

 Listening Suggestions refer you to audio examples that illustrate a concept or technique discussed in the text.

 Warnings caution you against conditions that may affect audio playback, impact system performance, alter data files, or interrupt hardware connections.

 Cross-References alert you to another section, book, or resource that provides additional information on the current topic.

 Online References provide links to online resources and downloads related to the current topic.

Chapter 1

Computer Concepts

...*What You Need from a Computer*...

In this chapter, we introduce you to the basic components of a computer system and the minimum configuration required to run Ableton Live software. We also discuss how to configure the hardware for your system and how to configure your computer's software settings for optimal results. Lastly, we take a brief look at how to work within your digital audio workstation, using Ableton Live as an example.

⊕ Learning Targets for This Chapter

- Understand how to select a computer for your digital audio workstation
- Understand how to navigate your computer's operating system for basic file management purposes
- Understand how to set preferences for your computer and your software
- Understand how to perform basic operations with your digital audio workstation

 Key topics from this chapter are illustrated in the Ableton Live Audio Production Basics Study Guide module available through the Elements|ED online learning platform. Sign up for a free account at ElementsED.com.

Selecting the right digital audio workstation (or DAW) involves many considerations: What features do you need? How big will your projects be? What kind of production work do you intend to do? What kind of system can you afford? And so on. Prior to making a purchase, you would be wise to check out a trial version or low-cost (feature-limited) edition of the product you are considering, if available.

Fortunately, for those considering Ableton Live as their DAW of choice, it is possible to run a fully-featured trial of Ableton Live Suite for 30 days. The trial mode has no limitations and is a great opportunity to get to know the basic features of the software.

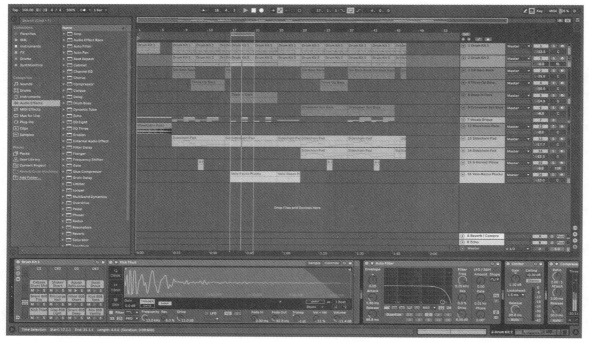

Figure 1.1 Ableton Live software

Before installing Ableton Live (or other DAW software), you'll need to verify that your computer system meets the minimum requirements for the software. Once you've cleared that hurdle and installed the software on a suitable computer, you will want to configure your computer and the software application for optimal performance. This chapter will help steer you in the right direction for these and other considerations to get a system up and running successfully.

Selecting a Computer

In this section, we will look at how to select a computer for use with Ableton Live or another digital audio workstation of your choice. We will also look at options for optimizing your computer setup to get the most out of your work environment.

Mac Versus Windows Considerations

One of your first considerations is whether to run your DAW on a Mac-based computer or a Windows-based computer. This decision is primarily one of personal preference, but it may also be based on the kind of system you already own.

If you are purchasing a new system for your DAW, understanding some general characteristics of Macs and Windows machines may help you decide on one platform versus the other.

Mac-Based Computers

One of the big selling points for Macs is how well they interact with other Apple products: iPhones, iPads, Apple TVs, and the Apple Watch. If you have already bought in to the Apple ecosystem, it may make sense to purchase a Mac computer for that reason alone. You will find the user experience on a Mac to be similar to that on other Apple products, in many respects.

Interoperability aside, Macs systems have a reputation for being easy to use and simple to understand. Additionally, Macs are known for high quality construction and attention to detail in their design.

The downside to purchasing a Mac is the higher price you will pay for those conveniences. You will likely spend more for a Mac-based system than you would for a similarly configured Windows-based system.

Windows-Based Computers

Windows-based computers are available from many different manufacturers, giving you an abundance of choices when selecting a system. The competition between manufacturers means it may be easier to find a system that meets your needs at a more affordable price point.

In addition to competitive pricing, Windows computers may also have an advantage in terms of the applications they support. If you use certain Windows-only applications for personal or work-related purposes, you may find that issue to be a deciding factor in your choice of platform.

Whichever platform you choose—Mac or Windows—be sure to purchase a system with adequate RAM, processing power, storage space, and connectivity options to support your needs.

The Importance of RAM

A computer's installed RAM, or random access memory, determines how much data and information can be stored in the computer's memory at any given time. RAM is generally used for temporary storage while an application is running. The RAM allocation is typically measured in Gigabytes (or GB).

Figure 1.2 Illustration of a typical RAM module

> **Running Windows on a Mac**
>
> If you prefer to work on a Mac but need to run Windows for certain applications, several options are available. Modern Mac computers allow you to run a Windows operating system (OS) on your Mac hardware. You can configure a Mac to run Windows 10, for example, using either a separate disk partition or a virtualization software option.
>
> To configure a separate disk partition for Windows 10, you can use the Boot Camp Assistant utility that comes with your Mac. This utility will partition your drive and guide you through the installation process for the Windows software. Using this option, you will need to restart your computer whenever you wish to switch between the Mac OS and the Windows OS.
>
> To run Windows 10 with virtualization software instead, you can purchase and install software such as VMware Fusion or Parallels Desktop. These applications allow you to run a Windows OS on your Mac desktop without rebooting.
>
> With either option, you will need to purchase Windows 10 separately and install it. Factor this in when considering the total budget for your system. The virtualization software (if used) and the Windows OS may also impact the amount of RAM, storage space, and processing power required for the system.

How RAM Is Used

When you select a paragraph of text in a word processor and choose the **EDIT > COPY** command, the text is stored temporarily in RAM. You can then move your cursor to a new location in the document and choose **EDIT > PASTE**. The text stored in RAM gets added to the document at the current cursor location.

Most applications also use RAM to keep track of the work you do in your open documents. As you work, a record of your changes and edits is saved in RAM. This allows you to use the **EDIT > UNDO** command, for example, to reverse any changes you've made.

The code required to run the application is also stored in RAM while the application is running. This allows the computer to work faster than it would if it had to run the application entirely from the hard drive.

Storing data in RAM allows the information to be retrieved and operated on very quickly.

However, once you quit an application or shut down your computer, the information stored in RAM is lost. That is why your changes must be saved to permanent storage first.

Why More RAM Is Better

DAWs such as Ableton Live rely on RAM to work efficiently. Like a word processor, a DAW uses RAM for edit operations. Your DAW will also use RAM to keep track of your changes and undoable actions.

In Ableton Live software, we call the list of user-undoable actions the *Undo History*.

The more RAM your computer system has installed, the more memory will be available to your DAW software for temporary storage. This means the application can run more efficiently, you can copy and paste larger amounts of data, and you can maintain a longer list of complex undoable actions.

Symptoms of insufficient RAM can include sluggish behavior, frequent error messages, long pauses (or hangs) when you perform certain actions, and sudden crashes.

DAW manufacturers typically publish minimum RAM specifications required to run the application. They may also suggest that you install more RAM for more efficient operation. For work on very large or complicated projects, you may want to exceed the manufacturer recommendations.

RAM Required for Ableton Live

The minimum RAM allocation to run Ableton Live on either a Mac or Windows computer is 4 GB, with 8 GB or more recommended. If you plan to run Ableton Live at the same time as other applications (iTunes, a web browser, an email client, etc.), you would be wise to target the high side of these numbers.

> **Installing More RAM**
>
> Depending on the model computer you buy, you may be able to add more RAM at a later time, as your needs grow. This can help keep the initial purchase price down. If this is your plan, make sure to do your homework before purchasing a computer. Certain models (especially "budget" options) have RAM chips soldered in place and cannot be upgraded later with more.
>
> Also note that upgrading RAM often requires you to replace all RAM in the computer, rather than simply adding to the existing RAM. Most computers have an even number of memory slots, with RAM assigned in pairs across them.
>
> For example, a laptop may have 4 GB of RAM allocated across two available memory slots. As such, each slot will have a 2 GB memory module installed. To upgrade to 8 GB of RAM, you would need to replace each module with a new 4 GB module. (RAM modules should be added in matching pairs.)

Processing Power

The processing power, or how fast a computer runs, is a function of the computer's central processing unit, or CPU. The CPU is responsible for performing all of the processing and calculations required by an application.

Figure 1.3 The computer's central processing unit determines how fast the computer can run.

Processing Speed. The speed of a CPU is based on its clock speed. Clock speeds for modern computers are measured in GHz (Gigahertz), or billions of cycles per second. A 1 GHz processor, for example, can process up to one billion instructions per second. The faster the CPU, the more powerful the computer, and the faster your applications will run.

Processor Cores. Another consideration is the number of processing cores in the computer chip. Many computer chips today offer dual cores, quad cores, or more. This means that the CPU includes multiple processing units within a single chip. The benefit of multiple cores is that the computer can process multiple instructions at one time. This in turn allows modern applications to run faster and more efficiently.

> ### Hyper-Threading
> Many CPUs support hyper-threading as an alternative (or supplement) to multiple physical cores. Intel's hyper-threading technology allows a single physical core to perform as two virtual cores. This has the same benefit as adding physical cores, as it doubles the number of instruction threads that the CPU can process at once when running an application.

How Processing Power Affects Your DAW

A host-based DAW like Ableton Live relies on the processing power of the host computer for all audio editing, mixing, and processing. A system with insufficient processing power may exhibit sluggish performance, unwanted audio artifacts, and frequent error messages. Projects involving large track counts and heavy use of plug-ins will be more susceptible to issues.

Certain audio processes are especially CPU-intensive. Some examples of common practices that may tax your CPU include the following:

- Using certain Edit operations that automatically split audio into many smaller pieces
- Stacking plug-ins on inserts, especially convolution reverbs and analog gear emulations
- Using many virtual instrument plug-ins in a project
- Using real-time Elastic Audio processing
- Configuring complex signal routing for submix groups, headphone outputs, and effects processing

To avoid running into processing limitations as your projects grow in complexity, look for a multi-core CPU that runs at a reasonably high speed.

Processing Requirements for Ableton Live

The minimum recommended CPU for Ableton Live is an Intel Core i5 processor running at 2.0 GHz or more. Most computer systems available today meet the minimum CPU requirements to run Ableton Live.

 Ableton Live supports multithreading, so you will get better performance out of a quad-core processor than you will out of a duo-core, for example.

> **Always Check Compatibility Before You Buy!**
>
> New versions of Ableton Live often have more stringent compatibility requirements than previous versions. Prior to buying a system, be sure to check the compatibility requirements on the Ableton website.
>
> Requirements for Ableton Live Intro, Standard, and Suite can be found at the following web addresses:
>
> - https://help.ableton.com/hc/en-us/articles/115001663530-Live-10-Minimum-System-Requirements
> - https://help.ableton.com/hc/en-us/articles/209775305-What-computer-should-I-buy-

Another important consideration when buying a computer is the amount of storage space the computer provides. This is a function of the computer's installed hard disk drive (HDD) or solid state drive (SSD). The size of the storage drive determines how many applications you can install and how much space you will have for saving files on your system.

Figure 1.4 Illustration of a hard disk drive

Determining How Much Drive Space You Need

When determining storage requirements for your DAW, you will need to consider multiple factors:

- How much space is required to install the DAW software application itself.
- How much space is required to install the plug-ins you intend to use with the DAW.
- How much space you will need for the projects and media files that you create with the DAW.
- How much space you will need for other media files that you may want to use with the DAW (such as sample libraries, loop libraries, and sound effects).

In addition, you may need to consider the impact of other files and applications that you want to use on your computer. For example, if you plan to store your digital photos and your personal music collection on the computer, you will need to allocate additional space for those purposes.

Likewise, if you will use the computer for work in programs such as Microsoft Word, PowerPoint, and Excel, you will need additional space to install each of those applications.

Installation Requirements for Ableton Live

To install Ableton Live software, you will need a minimum of 3 GB of storage space for the application and the included plug-in options.

 If you wish to install all of the additional content included with Ableton Live Suite you will need approximately 67 GB of available storage space.

Storage Options

Many storage options are available for today's computer systems. A computer's internal system drive may be either an HDD or an SSD. The internal storage can be supplemented by any number of external drives and storage options. To allow your files to be accessed from anywhere, you can also consider cloud-based storage.

HDD Versus SSD Storage

HDD. Hard disk drives use spinning metal disks to store data. The more disks in the drive, the higher the data capacity. Storage sizes for modern HDDs typically range from around 1 to 4 Terabytes (TB).

The data transfer speed for an HDD is based in part on the speed of the spinning disks, measured in revolutions per minute (or RPM). Common options include 5400 RPM drives and 7200 RPM drives.

Hard disk drives provide greater storage capacity at a lower price than their solid-state counterparts.

SSD. Solid state drives, by comparison, use flash memory for storage instead of spinning metal disks. The storage capacity of an SSD is based on the number and size of its memory chips. Storage sizes for SSDs typically range from 256 GB to 1 TB.

Solid state drives provide faster access speeds than HDDs, meaning they provide better read/write performance for the computer's functions. This is noticeable in faster start-up times for the computer, faster launch times for applications, and faster file transfers among systems and drives.

Solid state drives also require less power than HDDs. This gives them an advantage for laptop computers and mobile devices. Additionally, SSDs have no moving parts, so they operate silently and are less prone to mechanical failure. These advantages come at the expense of higher cost and lower overall storage capacity.

External Drive Storage

A popular option for supplementing a computer's storage is to use an additional, external drive. External drives are readily available, offering a variety of connection types to match the available ports on your computer. Some common connection types include the following:

- FireWire 400 or 800
- USB 2.0 or 3.0
- USB-C
- eSATA
- Thunderbolt 1, 2, or 3

If you plan to store media files on an external drive for use with your DAW, look for a high-performance drive with a high-speed data connection to your computer. Good drive options would include a 7200 RPM HDD or an SSD. Good data connections would include USB 3.0, USB-C, eSATA, Thunderbolt 2, or Thunderbolt 3. Be sure to match the capabilities of your computer when selecting the connection type.

Cloud Storage

Both the internal system drive of your computer and any connected external drives are considered local storage. Your local storage can be complemented by any number of cloud-based storage options.

Cloud storage utilizes large-capacity storage servers and data centers that you connect to via the Internet. Cloud storage options generally require a subscription. To allow for quick access to your cloud-based content, cloud storage services usually allow you to sync your cloud content to your local storage.

The advantages of using cloud storage are many. For example, cloud storage enables you to keep your files in sync across multiple devices, to easily share files with anyone who has an Internet connection, and to retrieve your files from any computer, anywhere in the world.

Some popular cloud-based storage services include Dropbox, SugarSync, and Google Drive.

Onboard Sound Options (Audio In and Out)

Another aspect to consider when selecting a computer is the onboard sound options that the model offers. Many DAWs, including Ableton Live, can utilize the computer's built-in sound options for recording and playback. This allows you to use any built-in microphone inputs on the computer for recording to your project and to use the built-in computer speakers for playing back your project.

While you may not want to use the computer's microphones for any meaningful recording project, having onboard audio will allow you to run your DAW without requiring a connected audio interface. This can save you some money at the outset. It also provides the flexibility to work on the go, especially when using a laptop, without needing to bring along additional hardware.

 Although audio inputs are not required, audio outputs are critical. Look for a computer that has built-in speakers and/or a stereo headphone jack for monitoring playback from your DAW.

Other Options to Consider

Aside from the computer hardware itself, you may also want to consider options for peripheral devices. With the right accessories, you can customize your setup for efficiency and/or personal preference.

Mice and Trackballs

Many computer users find the mouse that comes with their computer to be less than ideal for long-term use. Numerous alternatives are available, offering features such as multiple buttons, scroll wheels, trackballs, gesture/touch input, and wireless connection to the computer. Mouse upgrades can range from around $20 to upwards of $100. Be sure to try out any options you are considering while running your DAW (or a similar application) to determine what works best for you.

Trackpads and Touchpads

As an alternative or supplement to a traditional mouse, Apple offers several "Trackpad" options (such as the Magic Trackpad and Magic Trackpad 2). These devices provide touch-based gesture input, similar to that found on many modern laptops. Logitech, Dell, and others offer similar "Touchpad" devices for Windows computers.

> ### Benefits of an Audio Interface
>
> Many audio interfaces are available that are compatible with Ableton Live. Any audio interface that is Core Audio-compliant (Mac) or ASIO-compliant (Windows) will work with Ableton Live.
>
> Some good options include the **Focusrite Scarlett** series. These Focusrite products each come bundled with Ableton Live Lite, along with a number of additional plug-ins. Many Focusrite interfaces are also available in **Studio** editions, with microphone and headphones included.

> The benefits of using an audio interface can include better sound quality, a greater number of available inputs and outputs (I/O), and access to digital I/O options. For many beginning users, the number of inputs may be the most significant issue. Ableton Live Lite supports up to 8 simultaneous inputs, with a compatible interface, meaning that you can record from up to 8 different sources at once. Without an audio interface, you are limited to the input options available on your computer.

Working with Your Computer

Once you have selected an appropriate computer to host the DAW of your choice, you should spend some time to get familiar with basic operations on the system. If you already have experience with computers and the operating system you will be using, you may want to skip or skim this section. The information in this section focuses on navigating among the files, folders, and applications on your computer.

In this section, we cover basic operations such as locating, moving, and organizing files; creating folders; launching applications; and saving files.

File Management

Perhaps the most important thing you should know about your computer is how to navigate among the files and folders using the computer's operating system. On the Mac, you will use the Mac Finder for this purpose. On a Windows computer, you will use File Explorer windows.

Using the Finder (Mac-based OS)

The desktop experience in a Mac-based operating system is defined by the Finder application. The Finder is the first thing you see after you start up a Mac, before you launch any applications or open any files or folders.

Figure 1.5 The Finder icon on a Mac

The Finder includes a menu bar at the top of the screen, the desktop display in the center (background image along with any desktop icons and open Finder windows), and the Dock at the bottom of the screen. The term "Finder" is commonly used interchangeably with the term "desktop" on a Mac.

12 Chapter 1

Figure 1.6 The desktop in the Mac Finder (OS X 10.11 El Capitan)

> ### The Finder Versus the Desktop
>
> Technically speaking, the Finder and the desktop are two different things. The Finder is an application on the Mac that is always running. You can switch to the Finder application at any time, the same way you can switch between other running applications.
>
> The desktop is the default location displayed by the Finder. You can think of the desktop as the background canvas upon which your open folders and applications are displayed. The desktop also provides a location to store folders and files.
>
> The items stored on the desktop are visible from the Finder. They can also be displayed in a Finder window, by opening the Desktop folder.

From the Finder, you can:

- Navigate through the file system on your storage drive(s)
- Open folder locations, and
- Manage files within open folders

To open a new Finder window, choose **FILE > NEW FINDER WINDOW** from the menu bar at the top of the screen (or double-click on a folder on the desktop). You can navigate within an open Finder window by clicking on a location in the sidebar or by double-clicking on a displayed folder in the file list to open it.

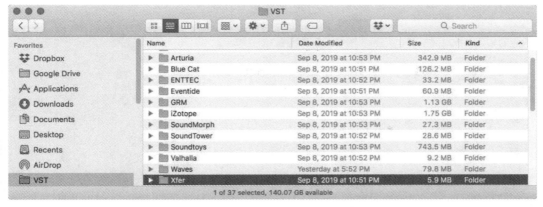

Figure 1.7 An open Finder window (Mac OS X)

Using File Explorer (Windows-based OS)

Windows computers also provide a desktop view (similar to the Mac Finder). However, the desktop might not be the first thing you see upon startup, depending on the operating system you are using. (Windows 8 and Windows 8.1 use the "Metro UI" tile view by default, similar to that on a Windows Phone.)

To open a new File Explorer window, do one of the following:

- In Windows 7 or Windows 10, click on the yellow File Explorer icon in the taskbar at the bottom of the desktop screen (or double-click any folder on the desktop).

Figure 1.8 Mouse cursor next to the File Explorer icon in the taskbar (Windows 7 shown)

- In Windows 8 or Windows 8.1, first click the **DESKTOP** tile from the tiled Start screen; then from the desktop, click the File Explorer icon in the taskbar (or double-click any folder on the desktop).

> (i) In Windows 8.1, you can also access the File Explorer directly from the tiled Start screen by typing *file explorer* and clicking on the File Explorer icon that appears.

As in the Mac Finder, you can navigate within an open File Explorer window by clicking on a location in the Navigation pane on the left or by double-clicking on a displayed folder in the file list to open it.

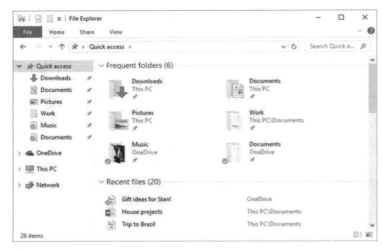

Figure 1.9 The File Explorer window (Windows 10)

Locating, Moving, and Opening Files

Using the locations in the sidebar of a Mac Finder window (or the Navigation pane in File Explorer), you can navigate to different folder locations on your computer. For example, clicking on the **DOWNLOADS** location will take you to the Downloads folder on your system (for any files you've downloaded from the Internet). Clicking on the **DOCUMENTS** location will take you to your Documents folder.

You can access the files within a folder at a given location by double-clicking on the folder to open it. On a Mac, you can also click on the triangle next to a folder to expand it and display its contents.

Figure 1.10 Finder window with the Ableton folder expanded

Moving Files. To move a file from one location to another, you can drag the file from its current location to any other visible folder within the window or to any location displayed in the sidebar/Navigation pane. You also can drag a file from one open Finder window or File Explorer window to another.

Opening Files. Opening a file from a Finder or File Explorer window is usually as simple as double-clicking on the file. In most cases, the file will open in the application that created it, if available on your computer. If the application is not available—or if your computer cannot determine an appropriate application to use for the file—you can choose a compatible application on your system. On both Mac- and Windows-based systems, you can right-click on a file and select an application to use for the file from the right-click menu.

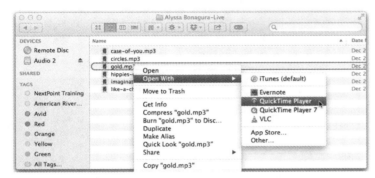

Figure 1.11 Right-click menu for a file in a Mac Finder window

---

Accessing Right-Click Functions on a Mac

For users who are new to the Mac, using the mouse can require an adjustment. The Apple mouse doesn't offer separate left- and right-click buttons. Instead, you must rock the mouse to get the desired effect. For a standard left-click, you must click on the *left side* of the mouse, causing the mouse to rock to the left. For a right-click, you need to click on the *right side*, causing it to rock to the right.

To improve your results, practice using two fingers on the mouse. Place your index finger on the left side and your middle finger on the right. (Reverse if you are left-handed.) For a left-click, roll your hand to the left and click with your index finger. For a right-click, roll to the right and click with your middle finger.

If you are using a Mac trackpad instead of a mouse, use a two-finger click to access right-click functions.

Creating Folders

On both Windows and Mac, you can easily create a new folder on the desktop or at another desired location. While displaying the desktop or viewing a location in a Finder or File Explorer window, simply right-click with the mouse and select **New Folder** (Mac) or choose **New > Folder** (Windows) using the pop-up menu. (See Figures 1.12 and 1.13.)

You can also create a new folder by using a menu command (Mac) or Ribbon button (Windows 8 and later) or by using associated keyboard shortcuts: **Command+Shift+N** (Mac) or **Ctrl+Shift+N** (Windows).

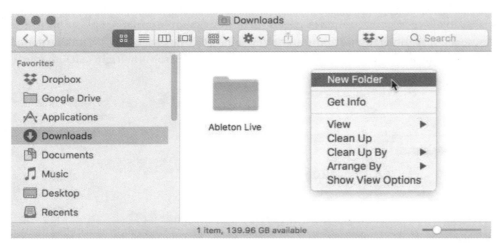

Figure 1.12 Creating a folder by right-clicking on a Mac

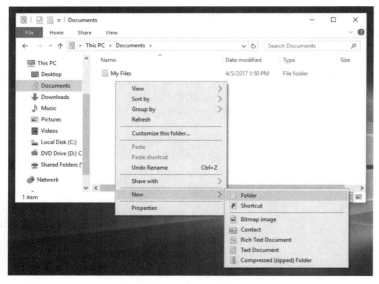

Figure 1.13 Creating a folder by right-clicking on Windows

Launching Applications

To launch an installed application, you can either double-click on the app file in your computer's Applications directory or use one of the shortcut operations provided by your OS.

Mac-Based Systems. The easiest path to opening an application on the Mac is the Dock. The Dock is a bar of icons that sits at the bottom of your Mac screen by default. It provides quick access to your installed apps. To launch an app, simply click on its icon in the Dock.

Figure 1.14 Launching Ableton Live 10 Lite from the Dock

Windows-Based Systems. One easy way to launch an application on any Windows system is to use the Search function. Simply press the **START** key (also known as the **WINDOWS** key) and begin typing the name of the application you wish to open. (See Figure 1.15.)

 Pressing the START key followed by any text activates a Start Menu search in Windows 7, a search in a right-hand pane in Windows 8, or a Cortana search in Windows 10.

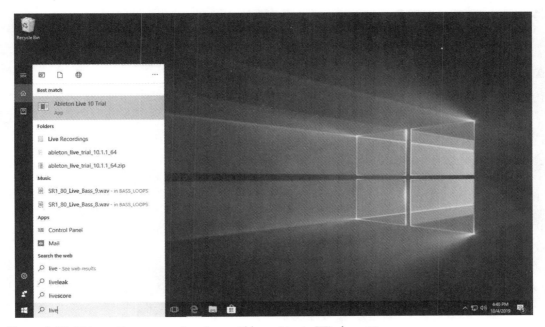

Figure 1.15 Using a Cortana search to locate Ableton Live in Windows 10

> ### Launching an App from the Desktop
>
> During the install process, many applications will offer the option to place a shortcut icon, or *alias*, on your desktop. This shortcut links to the application file in your Applications directory.
>
>
>
> **Figure 1.16** Ableton Live shortcut on the Windows 10 desktop
>
> Shortcut icons provide a convenient option for launching an application, as you can simply double-click on the shortcut from the desktop to start the application.

Saving Files

Ableton Live allows you to start working on a project and save it at a later point, once you've created something worth keeping. In Ableton Live, the **SAVE LIVE SET** command is used to save the current project. The first time you save a project, you will need to choose a name and save location for the project.

Figure 1.17 Selecting the Save Live Set command

> **The Ableton Live File Hierarchy**
>
> The concept of a file hierarchy is common to all varieties of Ableton Live software. This structure is used to allow the software to reference media files, rather than embedding them inside of the Ableton Live Set.
>
> For example, any audio files that you record or import into Ableton Live will be stored separately within the project hierarchy. The actual Ableton Live Set file simply links to the audio files, keeping the size of the Set file itself to a minimum.

Once you have created your initial project, you can save the changes you make as you work by using the standard **FILE > SAVE LIVE SET** command.

Working with an Application

With your application up and running, you will be able to begin working on a project—whether that be a slide presentation, a spreadsheet document, or an audio recording project. Most modern applications, including Ableton Live, use menus and keyboard shortcuts for accessing their main commands.

Menus

The term *menu* is commonly used in software to denote one of the pull-down lists of commands at the top of the application. On Windows computers, the menu bar is typically anchored to the top of the application window. On Mac computers, the menu bar is typically anchored to the top of the screen.

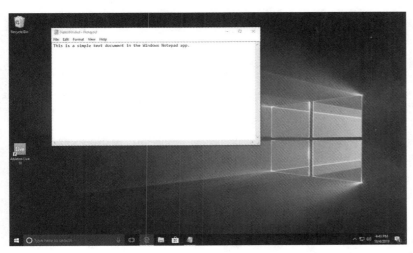

Figure 1.18 Menus in the Notepad app on Windows 10 (top of application window)

Figure 1.19 Menus for the TextEdit app on Mac OS X 10.11 (top of the screen)

To select a command, click on the associated menu name in the menu bar (such as the **FILE** menu) and then click on the desired command from the pull-down menu (such as the **SAVE** command). In some cases, menu items will include a submenu that you can use to select from between multiple choices.

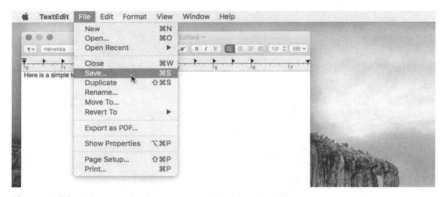

Figure 1.20 Selecting the Save command under the File menu

At times, you may have to distinguish between menus used by different applications or by the operating system (such as the Mac Finder). On a Windows-based system, you will likely find it easy to tell which menus belong to a given application, since the menus are anchored to the application window.

On a Mac-based system, however, you can run into situations where an application's windows are closed, but the application is still active. In these cases, you may see the desktop or another running application on screen, making it less apparent which application is active.

Figure 1.21 Safari menus displayed on a Mac with all application windows closed

To determine which application is active on a Mac, look for the application menu on the left side of the menu bar (immediately to the right of the Apple menu). The name of the active application will always display, regardless of whether the application has any windows open.

 Clicking on the desktop screen, an open Finder window, or a background app will make that item active, replacing the menus at the top of the screen.

 You can toggle through open applications by pressing COMMAND+TAB on a Mac or ALT+TAB on Windows.

Keyboard Shortcuts

As an alternative to using the menus to access commands, you may also be able to use keyboard shortcuts. Keyboard shortcuts use one or more modifier keys in combination with a standard key press to activate a command. Common examples include **COMMAND+S** on the Mac (**CTRL+S** on Windows) to save the open file or **COMMAND+C** on the Mac (**CTRL+C** on Windows) to copy a selection.

Mac and Windows systems each use different modifier keys, but their functions are similar. Table 1.1 shows the modifier keys available on each platform.

Table 1.1 Modifier keys available on Mac and Windows systems

Modifier Keys on Mac-based Computers	Modifier Keys on Windows-based Computers
The **COMMAND** key (⌘)	The **CTRL** key
The **OPTION** key (⌥)	The **ALT** key
The **CONTROL** key (^)	The **START** key (⊞)
The **SHIFT** key (⇧)	The **SHIFT** key

The advantages of using keyboard shortcuts include speed and efficiency, as well as reduced dependency on the mouse. This not only makes your work more precise, but it also helps avoid ergonomic problems, including carpal tunnel syndrome.

Often, keyboard shortcuts are listed next to the command names in the pull-down menus. You'll find it worth your time to memorize the shortcuts for some of the commands you use frequently in your DAW. Ableton Live includes numerous keyboard shortcuts that we will be covering in this book.

Review/Discussion Questions

1. What are some reasons for choosing a Mac computer for your DAW? What are some reasons for choosing a Windows computer instead? (See "Mac Versus Windows Considerations" beginning on page 3.)

2. How is RAM used by an application? Is RAM used for permanent storage or temporary storage? (See "How RAM Is Used" beginning on page 4.)

3. What are some indications that a computer system does not have enough installed RAM for the applications you are using? (See "Why More RAM Is Better" beginning on page 4.)

4. What factors affect how much processing power your computer has for running applications? (See "Processing Power" beginning on page 5.)

5. What are some examples of audio processes that can strain the CPU of an underpowered system? (See "How Processing Power Affects Your DAW" beginning on page 6.)

6. What are some things you should consider to determine how much storage space is required for your DAW computer system? (See "Determining How Much Drive Space You Need" beginning on page 8.)

7. What is the difference between an HDD and an SSD for storage? What are some advantages of each? (See "Storage Options" beginning on page 8.)

8. How are built-in sound options on a computer useful for Ableton Live? (See "Onboard Sound Options" beginning on page 10.)

9. How can you quickly navigate to different folder locations on your computer? (See "Locating, Moving, and Opening Files" beginning on page 14.)

10. What are some ways to create a new folder at the current location on your computer? (See "Creating Folders" beginning on page 15.)

11. How can you select a command in an application such as Ableton Live? (See "Menus" beginning on page 19.)

12. What modifier keys are available on Mac computers? What modifiers keys are available on Windows computers? (See "Keyboard Shortcuts" beginning on page 21.)

 To review additional material from this chapter and prepare for certification, see the Ableton Live Audio Production Basics Study Guide module available through the Elements|ED online learning platform at ElementsED.com.

Exercise 1

Exploring Audio on the Computer

🎧 Activity

In this exercise, you will download the media files used for this course and move them to an appropriate storage location on your system. Then you will use your computer to listen to sample files for an excerpt from a remix of the song "Overboard" by Sacramento-area band, The Pinder Brothers.

🕒 Duration

This exercise should take approximately 10 minutes to complete.

⊕ Goals/Targets

- Practice file management techniques
- Practice critical listening techniques
- Explore sonic elements that comprise a finished mix

Exercise Media

This exercise uses media files taken from the song, "Overboard," provided courtesy of Sacramento-area band The Pinder Brothers, along with additional files provided by Eric Kuehnl from a synthesizer-based remix of the song.

Written by: Matt Pinder; Performed by: The Pinder Brothers;
Produced by: Scott Reams and The Pinder Brothers*; Remix by: Eric Kuehnl and Frank D. Cook

The media provided for this course may be used for educational purposes only. No rights are granted to use the media for any other personal, commercial, or non-commercial purposes.

* The mix, processing, and media files have been adapted for use in the exercises contained herein.

Getting Started

To get started, you will download the media files for this course and locate them in your **Downloads** folder. You will then move the downloaded media files to your **Documents** folder or other appropriate location.

Download the course media files:

1. Launch the browser of your choice on your computer (Safari, Microsoft Edge, Firefox, Chrome, etc.)

2. Point your browser to www.halleonard.com/mylibrary and enter your access code (printed on the opening page of this book).

3. Click the **DOWNLOAD** link next to the Live APB Media Files listing in your My Library page. The Live ABP Media Files folder will begin downloading to your Downloads folder.

4. Open the Downloads folder on your system:

   - On Mac, activate the Finder and choose **GO > DOWNLOADS**.
   - On Windows, press **CTRL+J** from within your browser and then click the link labeled **OPEN FOLDER**, **SHOW IN FOLDER**, or similar.

Move the Live ABP Media Files folder to a permanent location:

1. Select the Live ABP Media Files folder in the Downloads folder and drag onto the Documents location displayed in the sidebar/Navigation pane of the Finder or File Explorer window.

2. Click on the Documents location in the sidebar/Navigation pane to open the Documents folder.

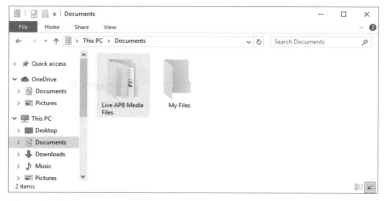

Figure 1.22 Live APB Media Files folder in the open Documents folder (Windows 10 shown)

Listening to Sample Files

In this part of the exercise, you will locate and play various audio files. In the process, you will practice listening for different musical parts in each file.

Listen to the music mix:

1. Double-click on the **Live APB Media Files** folder in the open Finder or File Explorer window to open the folder and display its contents.

2. Double-click on the **01. Pinder Brothers** folder to open it. You will see the following files in the folder:

   - StereoMix.wav
   - Stem1.wav
   - Stem2.wav
   - Stem3.wav
   - Stem4.wav

3. Select the **StereoMix.wav** file.

4. Do one of the following to listen to the selected file:

   - On a Mac-based system, press the spacebar to begin playback using the QuickLook feature.
   - On a Windows system, right-click on the file and select **Play with Windows Media Player**.

5. As you play back the mix, listen for each of the different instruments and sounds that you can hear.

   Write down each part that you hear in the table below. Rate the prominence of the part in the mix on a scale from 1 to 5, with 5 being very prominent/easy to hear and 1 being barely noticeable.

Instrument or Part	Prominence
Lead Vocal	5

Listen to the stem files:

1. Select the **Stem1.wav** file and listen to it using the same method that you used in Step 4 above.
2. Write down the main part you hear in the stem file, along with a short description, using the table below.
3. Listen for additional parts in the stem file and indicate each different part you hear in the table along with a short description of each.
4. Repeat the process using each of the other stem files (Stem2, Stem3, and Stem4).

   Not all of the stems will have different discernable parts. If you are not accustomed to picking out and describing the different musical sounds you hear, just do your best to make distinctions where something sounds different. This is good ear training regardless of what type of audio production you are interested in.

   Some terms you might use in your descriptions include the following: lead vocal, background vocal, vocal harmony, vocal double, guitar, synth pad, bass, kick drum (bass drum), snare drum, hi-hat, etc.

Stem File	Description of the Instrument or Part
Stem1	Lead vocal – Main vocal melody throughout

Finishing Up

After you've listened to each of the stem files, return to the **StereoMix.wav** file and listen through it again. Are you better able to distinguish each part in the mix, now that you've heard them in isolation?

You should now be able to clearly pick out things like the synth bass and vocal harmonies that you may have missed before. You may be able to distinguish subtler details as well. Can you hear where the guitar plays during the intro? Can you hear individual kick and snare drum hits? Can you tell where the hi-hat begins playing?

 To hear individual drum parts in isolation, check out the audio files in the z-Drum Parts folder. These files will help you learn what each drum piece sounds like and let you know what to listen for in a drum kit.

That completes this exercise. For more practice honing your listening skills, try identifying parts in popular songs that you hear on the radio or Spotify. If you are more interested in post-production applications, listen for different sound elements in your favorite TV program. Examples include dialog, background ambience (birds chirping, waves crashing), Foley (footsteps, character movements, noise from objects being handled), sound effects (telephones ringing, gun shots, laser beams), and background music.

Chapter 2

DAW Concepts

...*What You Need from Your Digital Audio Workstation*...

This chapter introduces you to the basic concepts of the DAW or Digital Audio Workstation. We begin by looking at some of the popular DAWs in use today. Then we explore the common audio plug-in formats that are used to add both effects and virtual instruments to a DAW. Next we look at the various Ableton Live editions that are available. We conclude this chapter with a brief explanation of some important terminology that you'll need to know to work effectively with Ableton Live.

Learning Targets for This Chapter

- Learn the basic concepts of Digital Audio Workstations
- Gain an understanding of common plug-in formats
- Explore available Ableton Live editions
- Become familiar with important audio terminology

Key topics from this chapter are illustrated in the Ableton Live Audio Production Basics Study Guide module available through the Elements|ED online learning platform. Sign up for a free account at ElementsED.com.

Although this book is written for Ableton Live, it is important for you as an audio production enthusiast to have at least a passing familiarity with other digital audio products on the market. Not only are you likely to encounter audio practitioners who favor these platforms, you may end up using some of them yourself from time to time, to supplement your work in Ableton Live.

Functions of a DAW

The term DAW stands for Digital Audio Workstation. This term is used generically to refer to any software that can be used to record, edit, and mix audio, although most modern systems can be used for MIDI recording and editing as well.

What Can a DAW Do?

Some DAWs excel at audio editing, while some are primarily known for their MIDI features, but almost every modern DAW can do both to some degree. Ableton Live is generally considered to be a superior platform for working with loops and MIDI content. However, Ableton Live also provides a robust platform for audio work.

In addition to basic recording and editing functions for audio and MIDI, DAWs often provide sound modules (or virtual instruments) for use with MIDI data. Such virtual instruments are commonly available in the form of plug-ins that use MIDI data to trigger sounds.

Most DAWs also use plug-ins for audio processing, to apply effects and polish to audio recordings. Additionally, DAWs typically provide mixing and automation functions, for blending tracks, creating an effective stereo image, and incorporating dynamic changes during playback.

Common DAWs

Many DAWs are available, ranging in price and feature set. Here, we provide an overview of some of the most popular DAWs and highlight some of the differences between them.

Ableton Live

Ableton's Live is a unique DAW that was created specifically as an electronic music performance tool. It offers a familiar track layout and audio and MIDI editing toolset in its Arrangement view. But where Live really stands out is the revolutionary Session view that took the DAW world by storm back in 2001.

In Session view, audio and MIDI clips are loaded into "slots" that can be triggered, stopped, and looped independently from one another. As a result, complex song arrangements can be performed on-the-fly in a way that isn't really possible with many other DAWs. This has made Live a favorite among live performers, DJs, and electronic musicians.

The inclusion of Cycling 74's MAX software with Ableton Suite (known as "Max for Live") offers another unique feature—the ability to create custom effects and virtual instrument plug-ins without writing any code.

In recent years, Live has probably seen more development of dedicated control surfaces than any other DAW, with Akai's APC series, Novation's Launchpad series, and Ableton's Push devices leading the way.

Figure 2.1 The Session View in Ableton Live

Avid Pro Tools

Pro Tools became the first commercially successful DAW in the audio world when it was introduced back in 1991. The original version of Pro Tools could play only four tracks of audio and cost around $6000 for the hardware and software.

Since then, Pro Tools has evolved from a tool built exclusively for audio professionals to the popular entry-level DAW that it is today. The Pro Tools audio editing feature set is still unsurpassed in its power and simplicity, making it the de facto standard in professional music production and audio post-production for film and TV. The introduction of sophisticated control surfaces in the late 1990s furthered its professional evolution, making Pro Tools an extremely powerful mixing tool.

In the 2000s, Pro Tools made great strides as a MIDI production tool, adding a number of advanced MIDI features as well as the ability to edit using standard music notation. Cross-integration with the Sibelius notation platform has made it easy to interchange MIDI information between Pro Tools and Sibelius, for sophisticated, professional-grade scoring.

Figure 2.2 A project in Pro Tools

Apple GarageBand

GarageBand is Apple's entry-level DAW. As an Apple product, it is available only for Mac OS computers and iOS devices, and it won't run on Windows computers or Android devices.

GarageBand makes it easy to get up and running to start making music right away. This DAW includes a nice complement of effects plug-ins including EQ, compression, reverb, and delay, as well as excellent guitar amp and pedal emulations.

The included virtual instrument plug-ins span the gamut from synthesizers to acoustic instruments to an automatic drum pattern generator. GarageBand also features a large selection of Apple Loops, which are professionally-recorded sound files that can be used to build a song very quickly.

The mobile version of GarageBand can run on both iPads and iPhones, making it easy for users to sketch out a song on-the-go, and then transfer the files to a Mac OS computer for further polishing. You can also import GarageBand songs directly into Apple's professional DAW, Logic Pro X.

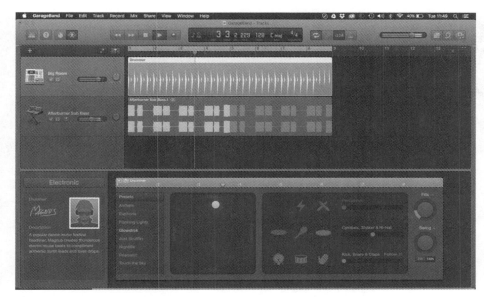

Figure 2.3 The Drummer instrument in Apple GarageBand

Apple Logic Pro X

Logic is GarageBand's big brother. Like GarageBand, Logic runs only on Mac OS and iOS platforms. The product was originally developed by a company called Emagic, which got its start creating MIDI sequencing software for Atari computers.

Logic includes the same core feature set as GarageBand, but it expands upon that to offer a truly professional audio production tool. Logic features a huge number of plug-ins effects including pro-level EQ, dynamics, and modulation effects. It also features a convolution reverb effect that can be used to create ultra-realistic models of real acoustic spaces.

The included virtual instruments in Logic are equally impressive. Logic includes a number of synthesizers, a physical modeling synth, a sampler, several drum instruments, as well as an organ and clavinet. Many MIDI composers and producers select Logic as their go-to DAW due to its robust support for virtual instruments.

Additional power-user features in Logic include Flex Time and Flex Pitch. These features allow users to freely adjust the timing and pitch of recorded audio.

Figure 2.4 Flex Pitch editing in Apple Logic Pro X

Steinberg Cubase/Nuendo

Steinberg's Cubase is one of the oldest and most respected DAWs on the market. The product began its development in the 1980s as a MIDI sequencer for the Atari ST (like Logic). Cubase has since evolved into a world-class DAW with outstanding MIDI capabilities and solid audio editing.

In particular, Cubase is very popular with hardcore MIDI users (such as film composers). This is due to several innovations that Steinberg introduced, including an ingenious chord track and elegant sampler articulation management. Additionally, the included VST plug-ins are top-notch.

 For information on VST and other plug-in formats, see "Plug-In Formats" later in this chapter.

Steinberg also manufactures another well-known DAW, by the name of Nuendo.

Nuendo is essentially an enhanced version of Cubase with more advanced audio features that target the audio post-production industry. Cubase and Nuendo both run on Mac OS and Windows computers, making either option an excellent choice for users who need to work on Windows or move between the two platforms.

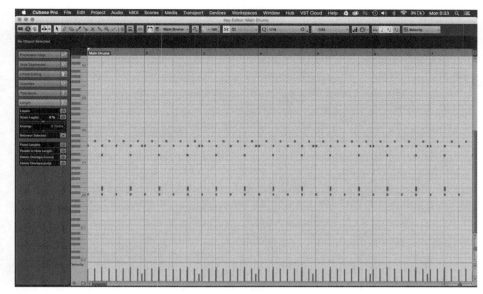

Figure 2.5 The powerful key editor in Steinberg Cubase

And the Rest…

Numerous other DAW options are available, with more coming to market everyday. Some examples of other popular DAWs include the following:

- **Cakewalk (formerly Sonar)**—Cakewalk by Bandlab, formerly known as Sonar, evolved from the original Windows-only Cakewalk application that was a popular early MIDI sequencer. It now offers powerful audio functionality and has recently crossed over to Mac OS.

- **Cockos REAPER**—REAPER is an inexpensive and flexible DAW. Many users are drawn to REAPER for its incredible customization options: REAPER allows users to create complex "actions" to accomplish almost any task. In addition, REAPER can be configured so that its editing features mimic the functionality of another DAW, making it easy to use for "switchers."

- **MOTU Digital Performer**—Digital Performer is another Mac-only DAW that has a loyal following in the film-scoring world. Its credentials include well-known users such as Danny Elfman. Digital Performer offers a number of editing tools that hold particular appeal for classically-trained composers and orchestrators, as well as a deep general toolset for MIDI and audio work.

- **PreSonus Studio One**—Studio One is a relatively young DAW that has garnered a lot of attention in recent years. It offers a nice balance of MIDI and audio tools and is constantly adding innovative new features.

- **Reason by Reason Studios (formerly Propellerhead)**—Reason was once an absolute necessity for any DAW user looking to add a cost-effective bundle of virtual instruments. Reason has since matured into a full-featured MIDI and audio editor. It continues to offer a huge range of virtual instruments with a user-friendly interface.

Plug-In Formats

Most DAWs support plug-ins for both audio processing and MIDI virtual instruments. But not all DAWs support plug-ins in the same format. As a result, a fair amount of confusion surrounds the issue of plug-ins and plug-in formats.

 Ableton Live uses the term "device" to describe Ableton's included instruments, audio effects, MIDI effects, and Max for Live devices. Ableton Live uses the term "plug-ins" to describe third-party plug-ins in VST and AU format.

What Is a Plug-In?

A plug-in is a small software program that runs inside of your DAW to extend its functionality. Plug-ins are typically inserted directly onto a track in the DAW to run in real-time.

Plug-ins fall into two general categories:

- **Effects Plug-ins**—Effects plug-ins offer a range of signal processing effects including equalization (EQ), dynamics processing (such as compressors and limiters), modulation (such as chorus and flange effects), harmonic processing (such as distortion), and time-based effects (such as reverb and delay).

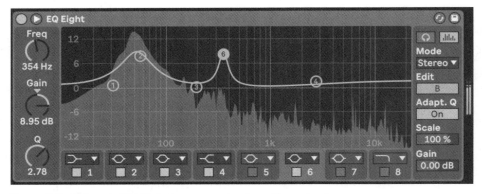

Figure 2.6 Ableton's EQ Eight is an example of an effects plug-in

- **Virtual Instrument Plug-ins**—Virtual instrument plug-ins are software emulations of hardware musical instruments. These instruments include synthesizers (both analog emulations and modern digital algorithms), samplers, and drum machines. (See Figure 2.7.)

Figure 2.7 Wavetable by Ableton is an example of a virtual instrument plug-in

Common Plug-In Formats

Some common plug-in formats include the VST format developed by Steinberg, the AU format developed by Apple, and the AAX format developed by Avid.

The following breakdown provides details on each of these formats:

- **VST**—Virtual Studio Technology (VST) is a plug-in format created by Steinberg. It was originally developed for Cubase in the 1990s, but the format was rapidly adopted in most of the other DAWs of that era. VST is easily the most popular format for third-party plug-ins and is supported in every major DAW except for Logic, GarageBand, and Pro Tools.

- **AU**—The Audio Units (AU) format is Apple's audio plug-in format. AU plug-ins are used primarily in Logic and GarageBand, but they are also supported in several other DAWs (including Ableton Live) as well as a number of simpler audio applications on Mac OS.

- **AAX**—The Avid Audio eXtensions (AAX) specification is the current plug-in format supported in Pro Tools. This specification supports both AAX Native, AAX DSP, and AAX AudioSuite formats. AAX plug-ins use 64-bit floating-point math and support sample rates up to 192 kHz. AAX plug-ins are not currently supported in any other (non-Avid) DAW.

Ableton Live Configurations

The requirements for your digital audio recording projects will determine the edition of Ableton Live that you will need to use. Ableton Live 10 is available in three commercial editions: Ableton Live Intro, Ableton Live Standard, and Ableton Live Suite.

Software Editions

Ableton Live Intro (and the similar Ableton Live Lite) is a basic version of Ableton Live, with limitations on editing features, track count, scene count, and I/O capabilities.

Ableton Live Standard provides the full editing feature set, unlimited tracks and scenes, and up to 256 channels of I/O. This book focuses on Ableton Live Standard software features.

Ableton Live Suite offers all of the Ableton Live Standard features, and adds additional virtual instruments, audio and MIDI effects, and Max for Live.

All Ableton Live editions are available for both Mac and Windows operating systems.

> ### Ableton Live Intro vs. Ableton Live Standard vs. Ableton Live Suite
> Throughout this book, we use the term **Ableton Live** to refer generically to all available software editions and the terms **Ableton Live Intro**, **Ableton Live Standard**, and **Ableton Live Suite** to refer to specific editions of Ableton Live software.

Table 2.1 provides a feature comparison of basic functionality for the three Ableton Live software editions.

Table 2.1 Comparison of Ableton Live software features

FEATURES	Ableton Live Intro	Ableton Live Standard	Ableton Live Suite
Software Instruments	4	5	15
Sounds	1500+ 5+ GB	1800+ 10+ GB	5000+ 70+ GB
Audio Effects	21	34	55
MIDI Effects	8	8	17
Supported Plug-In Formats	VST2, VST3, Audio Units	VST2, VST3, Audio Units	VST2, VST3, Audio Units
Audio Resolution	32-bit/192 kHz	32-bit/192 kHz	32-bit/192 kHz
Audio & MIDI Tracks	16	Unlimited	Unlimited
Scenes	8	Unlimited	Unlimited
Send and Return Tracks	2	12	12
Audio Inputs	8	256	256
Audio Outputs	8	256	256
Max for Live	No	No	Yes

 Ableton Live Lite is a feature-limited edition of Live that can be found bundled with other commercial audio products. Live Lite comes with a limited set of sounds and effects. Live Lite is further restricted to 8 Audio and MIDI tracks.

Audio Interface Options

All editions of Ableton Live software support a range of audio hardware:

- **Basic Interfaces** – Ableton Live supports a variety of hardware interface options on both Mac and Windows computers. These include your computer's built-in audio hardware or an external interface that connects via USB, FireWire, or Thunderbolt. Many of these peripherals can be powered by the computer's USB bus, enabling portability for laptop systems.

Figure 2.8 Focusrite's Scarlett 2i2 is an excellent entry-level interface

- **Mid-Range Interfaces** – Ableton Live Standard and Ableton Live Suite support up to a maximum of 256 channels of input and output, so much larger interfaces (or combinations of interfaces) can be used. You should consider a mid-range interface if you need more than just a couple of channels of input and output. You should also consider stepping up to a mid-range interface if you're looking for better quality microphone preamps and analog-to-digital and digital-to-analog converters.

Figure 2.9 MOTU's UltraLite Mk4 is an affordable, mid-range interface

- **High-End Interfaces** – If you're looking for an interface with a large number of inputs and outputs, you'll most likely want to consider a high-end interface from manufacturers such as Antelope Audio, Apogee, Focusrite, or Universal Audio. These interfaces offer up to 64 channels of input and output to accommodate the largest recording configurations. You may also want a high-end interface if you're looking for the absolute best quality in terms of microphone preamps and convertors.

Figure 2.10 Antelope Audio's Orion Studio is a high-end interface with 32 channels of I/O

Registering Your Software

Before you can install Ableton Live, you'll need to create an account on Ableton.com, and enter your product registration code. If you purchased a boxed version of Ableton Live or Push, your registration code will be included in the box. If you purchased a digital download of Ableton Live from a reseller, your registration code can be found in your confirmation email.

 If you purchased live through Ableton.com, the software will already be registered to your account.

If you've never registered an Ableton product, do the following:

1. Go to Ableton.com and create an account. You will be taken directly to the product registration page.

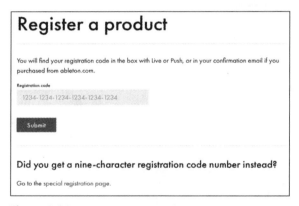

Figure 2.11 The product registration page on Ableton.com

2. Enter your Registration code and click the **SUBMIT** button.

If you've previously registered an Ableton product, do the following:

1. Go to Ableton.com and log in to your account.
2. Click on the **ACCOUNT** link in the upper right corner of the webpage.
3. Click on the pop-up menu next to "Licenses" and choose **REGISTER A PRODUCT**. (See Figure 2.12.)

Figure 2.12 Choosing Register a product from the Licenses menu

You'll be taken to the registration page, where you can register your new or upgraded product.

Software Installation and Operation

Separate installers are available for each version and edition of Ableton Live. Each Ableton Live installer also installs a variety of included software devices providing additional functionality. Supplemental devices and content (known as "packs") may be included as well, but require separate installation through the Ableton website or directly within the Ableton Live application (Ableton Live 10 only).

 Ableton uses the term "version" to differentiate between numbered software releases, such as Live 9 or Live 10. Ableton uses the term "edition" to differentiate between the flavors of Live, such as Intro, Standard, and Suite.

To access the installers for Ableton Live, do the following:

1. Go to Ableton.com and log in to your account.

2. Click on the **ACCOUNT** link in the upper right corner of the webpage. The Account page will appear.

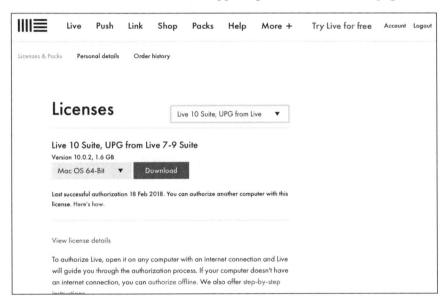

Figure 2.13 The Account page on Ableton.com

3. Generally, the most recent Ableton Live installer is shown by default. If necessary, you can click on the pop-up menu next to **Licenses** and choose an alternate version of Ableton Live.

4. Choose the appropriate operating system for your computer (Mac OS or Windows), and click on the Download button to download the installer. (See Figure 2.14.)

Figure 2.14 Viewing the available installers in your account

Included Devices and Packs

Devices are special-purpose software components that provide additional signal processing and other functionality within Ableton Live. Ableton devices come in three varieties: instruments, audio effects, and MIDI effects.

 If you're familiar with other DAWs, such as Pro Tools, Logic, or Cubase, you probably know the term "plug-in." Ableton use the term "devices" to describe the plug-ins included with Live. Ableton use the term "plug-in" to describe optional third-party plug-ins in either VST (Mac and Windows) or AU (Mac-only) format.

Ableton Live Intro and Lite editions come with a small set of Ableton devices. Ableton Live Standard adds one more instrument and most of the audio effects. Ableton Live Suite comes with the full complement of Ableton devices and also includes Max for Live.

Audio and MIDI Effects Devices

Ableton audio effects devices include a standard assortment of EQ, dynamics, reverb, delay, modulation and harmonic effects, and many more specialized devices that can be used to create unique and exotic results. While Live Intro offers just the essentials, there are more than 46 audio effects in Ableton Suite!

Ableton MIDI effects devices are also quite varied and include arpeggiation, chord creation, velocity and pitch manipulation, and several more.

Instrument Devices

Ableton instrument devices are a collection of virtual instruments that use a MIDI input to create sound. Live Intro offers a basic set of instruments including percussion devices (Drum Rack and Impulse), a basic sampler (Simpler), and a device for organizing and controlling instruments (Instrument Rack). Live Standard offers the same set but adds the External Instrument device that can be used to integrate hardware synthesizers or drum machines into your setup.

Live Suite incudes all of Ableton's instrument devices. In addition to those mentioned above, this set includes an analog synth (Analog), an FM and additive synth (Operator), a wavetable synth (Wavetable), a full-featured sampler (Sampler), and many more.

Installing Additional Packs

Depending on the edition of Ableton Live that you have purchased, you may be entitled to additional devices and content that were not automatically installed. Ableton builds these additional items into "Packs" that can be installed manually or through the Ableton Live application (Ableton Live 10 and later).

Installing Packs Manually

For all versions of Ableton Live, you can go to Ableton.com and download packs to manually install.

To access the installers for Ableton Live packs, do the following:

1. Go to Ableton.com and log in to your account.
2. Click on the **ACCOUNT** link in the upper right corner of the webpage.
3. Scroll down past the Ableton Live application installers until you see Packs. Click the link to the right to see all available packs (**ALL**) or only the packs that you have not previously downloaded (**NOT YET DOWNLOADED**).

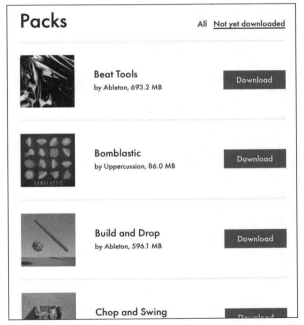

Figure 2.15 Viewing available packs in your Ableton account

4. Click on the **DOWNLOAD** link to download a specific pack.
5. Once the pack is downloaded, simply double-click on the installer and the Ableton Live application will automatically install the pack on your computer.

Installing Packs in the Ableton Live Application

Users of Ableton Live 10 and later have the ability to download and install packs directly from inside the Ableton Live application.

To download and install packs inside the Ableton Live application:

1. Verify that the **SHOW DOWNLOADABLE PACKS** preference is enabled in the Library tab of the Preferences window (**LIVE > PREFERENCES** [Mac] or **OPTIONS > PREFERENCES** [Windows]).

2. Click on **PACKS** under the Places section in the browser window. The browser will switch to the Packs view.

3. Click the triangle next to Available Packs to see the packs that are available in your account.

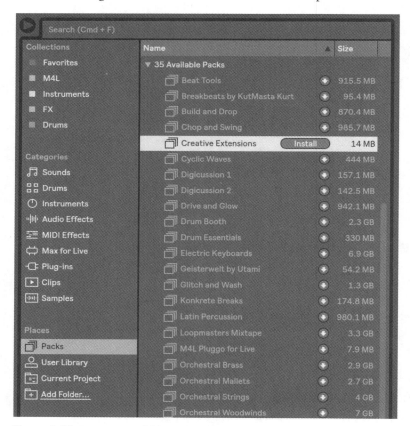

Figure 2.16 Viewing available packs in the Ableton Live application

4. Click the download arrow next to the desired pack. The pack installer will begin downloading to your computer.

5. Once the pack has downloaded, the Install button will appear. Click this button to automatically install the pack on your computer.

Launching Ableton Live for the First Time

The first time you launch Ableton Live, you will be asked to authorize the application. Once you've authorized the application, you may want to configure other options such as the audio interface settings.

Launching Ableton Live

Ableton Live software can be launched by double-clicking on the application icon on the system's internal drive or by double-clicking on a shortcut to the application. On Windows systems, the Ableton Live application is typically installed under C:\Program Files\Ableton\Ableton Live\AbletonLive10.exe, and a shortcut is placed on the desktop. On the Mac, Ableton Live is typically placed under Applications\Ableton Live 10.app and an application shortcut is placed in the dock.

Figure 2.17 The Ableton Live application icon

 On Windows systems, Ableton Live may also be available from the Start menu at the lower-left corner of the display.

When you first launch Ableton Live, the application will prompt you for authorization.

Authorizing Live

Ableton Live software uses copy protection, as do many other software products and plug-ins. The Ableton Live license is intended for a single user. However, Ableton permits you to install the Live application on up to two computers belonging to the registered user.

The first time you launch Ableton Live an authorization dialog box will appear. (See Figure 2.18.) At this point, you must choose one of three options:

- Authorize with Ableton.com
- No Internet on this computer
- Authorize later

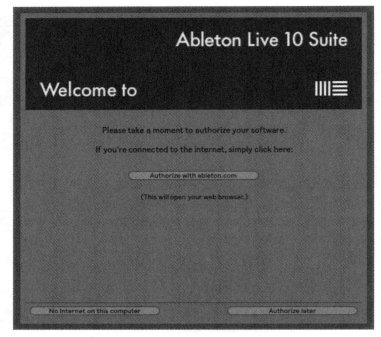

Figure 2.18 Live's authorization dialog box

Authorize with Ableton.com

If your computer has Internet access, you can quickly authorize Ableton Live with Ableton.com.

To authorize Ableton Live with Ableton.com, do the following:

1. Choose *Authorize with Ableton.com* from the authorization dialog box; your computer will automatically open a browser and take you to the Ableton website.

2. Log into your Ableton.com account. You'll be asked to select the license to authorize.

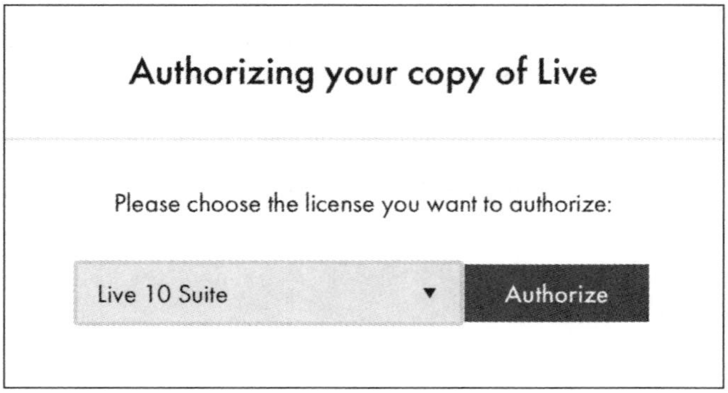

Figure 2.19 Selecting the license to authorize from your Ableton.com account

3. Assuming you have already registered Ableton Live using a registration code (see "Registering Your Software" above), choose the appropriate license from the Licenses list and click the **AUTHORIZE** button. Ableton.com will then send the authorization to your computer. This may take few seconds, during which time you will see a progress bar in the authorization dialog.

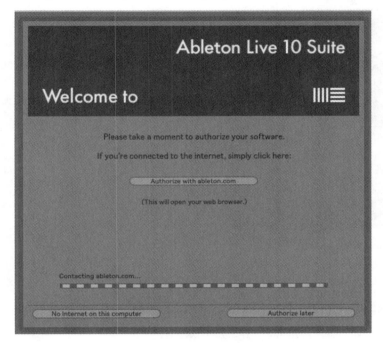

Figure 2.20 The progress bar in the authorization dialog

4. Once the authorization is complete you will get a confirmation message. Click **OK** to begin using Ableton Live!

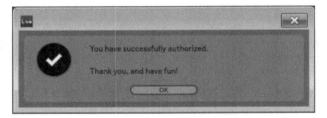

Figure 2.21 The authorization confirmation message

Authorizing Offline

If you have no Internet access on your computer or have difficulty authoring using the Authorize with Albeton.com option, you can authorize offline.

To authorize Ableton Live offline, do the following:

1. Choose **No Internet on this Computer** in the authorization dialog box. The offline authorization dialog will appear.

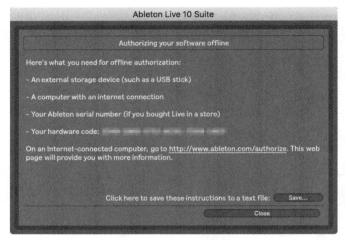

Figure 2.22 The offline authorization dialog box

2. Note the hardware code or save it to a text file by clicking the **Save** button.
3. On a computer with Internet access, log into your Ableton.com account.
4. Click the **Account** link in the upper right corner of the website.
5. Select the appropriate license from the Licenses list and then click the **Authorize Offline** link. This will take you to the Authorize Offline page.

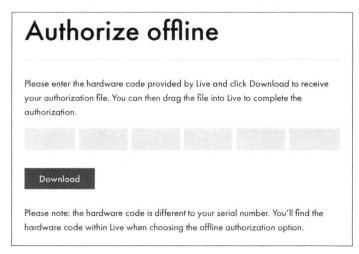

Figure 2.23 The Authorize offline page on Ableton.com

6. Enter your hardware code and click the **DOWNLOAD** button. Ableton.com will generate an authorization file (.auz), which will download to your computer.

7. Transfer the .auz file to the computer that you wish to authorize.

8. Double-click the .auz file (or drag and drop it onto the authorization dialog) to complete the authorization process.

Authorize Later

If you choose the **AUTHORIZE LATER** option from the authorization dialog, Ableton Live will launch but saving and exporting will be disabled.

Accessing Connected Audio Devices

When Ableton Live launches, it will generally be configured to use the default audio device on your computer. If no supported audio device is found, Ableton Live will launch with the audio engine off. You can immediately recognize that the audio engine is off because the CPU Load Meter at the far right of the toolbar will turn red and display the word "OFF."

Figure 2.24 The CPU Load Meter when the Audio Engine is off

To configure Ableton Live to use a specific audio device, such as a connected USB microphone or a separate audio interface, you will need to go to the **Audio** tab of the Live Preferences window. (See Figure 2.25.)

On Mac systems, choose **LIVE > PREFERENCES**; on Windows systems, choose **OPTIONS > PREFERENCES**. Then click on the **Audio** tab along the right side of the window. The audio input and output device selectors are displayed at the top of the window.

 If your supported audio interface is connected but not recognized by Ableton Live, you may need to install or update the device drivers. Check the manufacturer's website for the latest drivers for your interface.

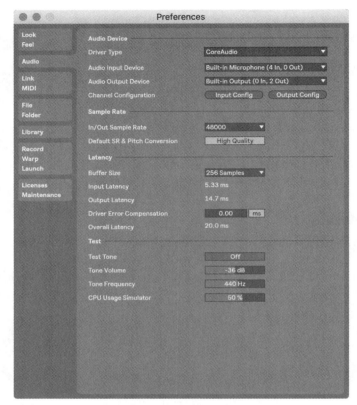

Figure 2.25 The Audio tab of the Live Preferences window

Optimizing Ableton Live Performance

The Audio tab of the Live Preferences window is also used to optimize the performance of the software. Like most audio software, Ableton Live utilizes the computer's processing capacity to carry out operations such as recording, playback, mixing, and plug-in processing (virtual instruments and effects). You can view the current CPU usage at any time by viewing the CPU Load Meter in the upper right corner of the toolbar.

Figure 2.26 The CPU Load Meter showing 7%

While the default system settings are adequate in some scenarios, Ableton Live lets you adjust a system's performance using the Sample Rate and Latency settings in the Audio tab of the Preferences window.

Sample Rate Settings

The Sample Rate setting will affect the quality of the audio that is going in or out of a connected audio interface. The Sample Rate setting will also determine the sample rate of audio files that result from recording on an audio track.

A high sample rate requires more processing power than a lower sample rate, which will reduce the number of tracks and plug-ins you can run on a particular computer. For this reason, many producers continue to work with sample rates of 44,100 and 48,000. While high-definition sample rates (such as 88,200 and 96,000) will result in better fidelity in the studio, they don't necessarily translate to better quality in final mixes that are distributed as MP3 files or streamed on services such as SoundCloud or Spotify.

 Refer to the section on "Moving Audio from Analog to Digital" in Chapter 3 for more information about how sample rate affects audio quality.

Latency Settings

The Buffer Size setting in the Audio tab of the Preferences window controls the size of the hardware buffer. This buffer handles processing tasks, such as plug-in processing.

- Low Buffer Size settings reduce monitoring latency when recording or monitoring an active input.

- High Buffer Size settings provide more processing power at the cost of higher monitoring latency.

As a general rule, the Buffer Size should be set as low as your project will allow while tracking, in order to minimize latency when monitoring an active input. You may need to use a higher setting for general playback purposes while editing and mixing, especially as your project becomes more complex.

Reducing Latency on Tracks with an Active Input

Ableton Live offers an additional option for reducing latency on tracks with an active input. By activating **OPTIONS > REDUCED LATENCY WHEN MONITORING**, any track whose input is set to **IN** or **AUTO** will have the smallest possible amount of latency given the current configuration of the Audio Engine.

Modifying Performance Settings

Adjustments to the Sample Rate and Latency settings can be made in the Audio tab of the Preferences window, as follows:

1. Choose **LIVE > PREFERENCES** (Mac) or **OPTIONS > PREFERENCES** (Windows).

2. Select the **AUDIO** tab.

3. From the Sample Rate section, click the **IN/OUT SAMPLE RATE** pop-up menu and choose the desired sample rate.

4. From the Latency section, click the **BUFFER SIZE** pop-up menu and select the buffer size in samples—lower the setting to reduce latency; raise it to increase processing power for plug-ins.

5. Close the Preferences window when finished.

Important Concepts in Ableton Live

There are a few key DAW concepts that you should master before getting started using Ableton Live. These include how the software handles audio files for your projects and the fundamental differences between audio and MIDI data. It will also be helpful to understand how to control playback and change the playback position in Ableton Live.

Set Files Versus Audio Files

Ableton Live manages two main file types: Set files and audio files. It is important to recognize the relationship between these file types.

- **Set Files**—Each time you begin a new project in Ableton Live, you will begin from an empty Set file. An Ableton Live Set file contains all of the information about the current audio production. This information includes the number and type of tracks in the project, the position of audio and MIDI data on each track, the device and plug-in assignments used on each track, the mixer level settings, the preferences, and a whole lot more.

    You will typically create a unique Set file for each song or production you work on.

- **Audio Files**—An audio file represents a single audio recording (or "take") that you've either created directly in your Set or imported into the Set from a disk location. A typical Set may have dozens or even hundreds of audio files associated with it.

    Ableton Live stores recorded or imported audio files in the Samples folder within the current project folder.

    Ableton Live stores audio files as WAV or AIFF files, but many other types of audio files can be imported into your project including FLAC, OGG Vorbis, MP3, and more. All audio files are referenced by the Set file rather than being embedded within it.

Audio Versus MIDI

The differences between audio and MIDI data are very important to understand when you're just starting out in the world of audio production. Audio files represent the audio signal that was recorded when the sound (from a voice, instrument, or other source) was captured using an audio interface. Therefore, the audio file itself actually *contains* the sound information that was captured. Audio files can be played back through the audio interface, allowing the sound to be heard using monitor speakers or headphones.

On the other hand, MIDI data is a series of note and control messages, typically recorded from a MIDI controller (like a keyboard or drum pads). (MIDI data can also be manually entered in Ableton Live.) The resulting MIDI file does NOT actually contain any sound information. The MIDI file's note and control messages must be sent to a real or virtual instrument, which can then turn the messages into an audio signal.

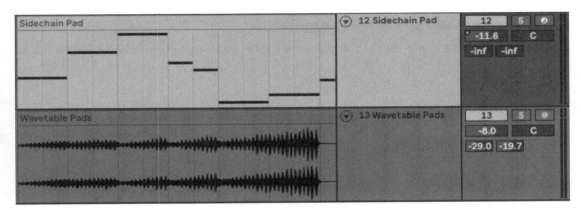

Figure 2.27 A MIDI track with a MIDI clip (top) and an Audio track with an audio clip (bottom)

Controlling Playback in Live

The **Play** button (solid triangle) in the Control Bar at the top of the main Live window serves to activate playback. Playback will begin at the Insert Marker location in the Arrangement View. The **Stop** button (solid square) serves to stop playback. Starting and stopping playback does not change the Insert Marker location. This means you can restart playback repeatedly from the same starting position.

Figure 2.28 The Control Bar showing the Play, Stop, and Arrangement Record buttons in the center

 You can also start and stop playback from the computer keyboard by pressing the spacebar.

Positioning the Insert Marker

You can reposition the Insert Marker as you work to change the playback location within the Arrangement View.

To play back a different portion of your Live Set, follow these steps:

1. Click anywhere within an existing track (avoiding clicking in the top half of a clip) to position the Insert Marker.

2. Press the **SPACEBAR** to begin playback from this point.

3. To stop playback, press the **SPACEBAR** again.

To move to a different playback point in the track, click a new position and press the **SPACEBAR** again.

Returning the Insert Marker to the Project Start

With the transport stopped, you can press the **STOP** button in the transport section of the Control Bar to move the Insert Marker back to the beginning of the project. This will allow you to start playback from the start of your Set.

Review/Discussion Questions

1. What are some tasks that can be completed using a Digital Audio Workstation (DAW)? (See "Functions of a DAW" beginning on page 32.)

2. What are some of the differences between the common DAWs? How are they similar? (See "Common DAWs" beginning on page 32.)

3. What is an audio plug-in? How are plug-ins enabled in a DAW? (See "What Is a Plug-In" beginning on page 38.)

4. Which of the common plug-in formats is/are supported in Ableton Live? (See "What Is a Plug-In" beginning on page 38.)

5. What are the three editions of Ableton Live software? (See "Ableton Live Configurations" beginning on page 39.)

6. What are some of the differences between Ableton Live Intro and the other editions of Ableton Live software? (See "Ableton Live Configurations" beginning on page 39.)

7. What is the maximum number of audio input and output channels supported in Ableton Live Intro? What about Ableton Live Standard and Suite? (See "Ableton Live Configurations" beginning on page 39.)

8. What are some examples of audio interfaces that are supported in Ableton Live? (See "Audio Interface Options" beginning on page 41.)

9. What are some of the settings you may need to specify upon first launching Ableton Live? (See "Launching Ableton Live for the First Time" beginning on page 47.)

58   Chapter 2

10. What are the two primary types of files that Ableton Live manages? (See "Important Concepts in Ableton Live" beginning on page 54.)

11. What are some of the differences between audio and MIDI information? (See "Audio Versus MIDI" beginning on page 54.)

 To review additional material from this chapter and prepare for certification, see the Ableton Live Audio Production Basics Study Guide module available through the Elements|ED online learning platform at ElementsED.com.

Exercise 2

Setting Up a Multi-Track Project

🎧 Activity

In this exercise, you will open a simple project containing the main parts for a remix of the song "Overboard" by Sacramento-area band, The Pinder Brothers. You will then configure the project settings for playback and play through the project.

🕓 Duration

This exercise should take approximately 10 minutes to complete.

◈ Goals/Targets

- Launch Ableton Live
- Open the Exercise Set
- Configure the Audio Preferences for playback and set the project tempo
- Save your work for future use

Exercise Media

This exercise uses media files from the song, "Overboard," provided courtesy of Sacramento-area band The Pinder Brothers, with additional files provided by Eric Kuehnl from a synthesizer-based remix of the song.

Written by: Matt Pinder; Performed by: The Pinder Brothers;
Produced by: Scott Reams and The Pinder Brothers*; Remix by: Eric Kuehnl and Frank D. Cook

The media provided for this course may be used for educational purposes only. No rights are granted to use the media for any other personal, commercial, or non-commercial purposes.

* *The mix, processing, and media files have been adapted for use in the exercises contained herein.*

Getting Started

To get started, you will launch Ableton Live as described in Chapter 1. Then you will open the exercise Ableton Live Set.

 Before starting this exercise, Ableton Live software (Intro/Lite, Standard, or Suite) must be installed and authorized on your system. If necessary, complete the Ableton Live download and installation steps, as outlined in Chapter 2.

Launch Ableton Live:

1. Launch Ableton Live using one of the following options (or another method of your choice):
   - On a Mac-based system, click on the Ableton Live icon in the Dock.
   - On a Windows-based system, double-click on the Ableton Live shortcut on the Desktop.

   A new Ableton Live Set will be created and will display on screen.

2. Choose **FILE > OPEN LIVE SET**. The Open Document dialog box will appear.

3. Navigate to your **Documents** folder or other location where you saved your **Live APB Media Files** folder in Exercise 1.

4. Double-click to open the **Live APB Media Files** folder, and then open the 02. **Starter Project** folder.

5. Locate the Overboard01.als Set file, select the Set, and click **OPEN**. The Set file will open and display as it was last saved.

 The Set file used for this exercise is compatible with all editions of Ableton Live (Intro/Lite, Standard, and Suite).

 If one or more samples are missing (offline), the Missing Files warning will appear in red at the bottom of the Ableton Live window. Click the warning message to resolve the issue.

 See Appendix A for details on locating missing files.

6. Choose **FILE > SAVE LIVE SET AS**. The Save Live Set As dialog box will appear.

7. Name your Set Overboard01-*xxx*, where *xxx* is your initials and navigate to an appropriate save location on your system. (See Figure 2.29.)

In a classroom or lab environment, you may be required to save or submit your work using a specific designated location. Check with your instructor for details.

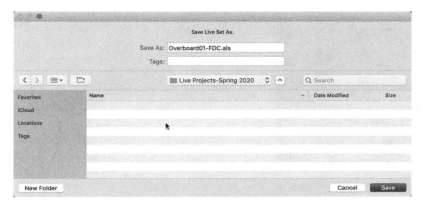

Figure 2.29 The Save Live Set As dialog box configured for this exercise

8. Click the **SAVE** button at the bottom of the Save Live Set As dialog to save the Set.

Playing the Set

In this part of the exercise, you will configure the Ableton Live preferences and play back the Set.

Configure the Live Preferences:

1. Choose **LIVE > PREFERENCES** (Mac) or **OPTIONS > PREFERENCES** (Windows) to display the Preferences window.

2. Click the **AUDIO** tab on the left side to show the audio preferences.

3. Verify that your connected audio interface (or built-in output option) is displayed in the Audio Output Device pop-up menu. If needed, click on the pop-up menus to select the desired options.

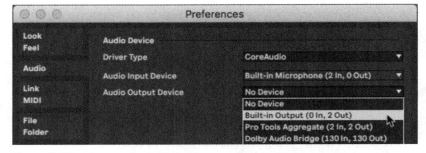

Figure 2.30 Selecting the Built-In Output as the Audio Output Device

62 Exercise 2

> ⓘ You may need to change the Driver Type before selecting the Audio Input and/or Audio Output Devices.

4. In the Latency section of the audio preferences, set the Buffer Size to at least **1024 SAMPLES**.

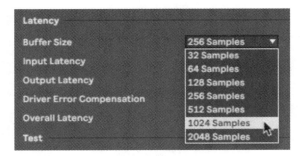

Figure 2.31 Setting the Buffer Size to 1024 samples

5. Close the Preferences window when finished.

Configure the Ableton Live windows:

1. Resize or maximize the Ableton Live window as needed for a full-screen view.

> ⓘ On current Mac systems, hold the Option modifier while clicking the green Maximize button at the top left of a window to maximize the window without obscuring the application menus at the top of the screen.

2. If necessary, choose **VIEW > TOGGLE ARRANGEMENT/SESSION VIEW** to make the Arrangement View visible, as shown below.

Figure 2.32 The Arrangement View

3. Click in the Tempo field in the top left corner of the Arrangement View so that it becomes highlighted (active).

Figure 2.33 Clicking in the Tempo field (shown active)

4. Enter a tempo value of 115.95. Press **ENTER** or **RETURN** when finished to accept the change; then press **ESCAPE** to deactivate the Tempo field.

Play back the Live Set

1. With the transport stopped, press the **STOP** button (solid square) in the transport section of the Control Bar to position the Insert Marker at the beginning of the project.

2. Press the **SPACEBAR** to begin playback. The playback cursor will scroll across the screen and you will hear playback (if you have speakers or headphones connected to your system).

3. While playing back the project, scroll the Arrangement window vertically, as needed, to view the audio waveforms on each track. Use the scroll bar on the right side of the window (or another method of your choice) to scroll up/down.

4. When finished, press the **SPACEBAR** a second time to stop playback.

Finishing Up

To complete this exercise, you will need to save your work and close the Set. You will be reusing this Set in Exercise 5, so it's important to save and retain the work you've done.

Finish your work:

1. Choose **FILE > SAVE LIVE SET** to save the Set.

2. If desired, you can exit Ableton Live by doing one of the following:
   - On a Mac-based system, choose **LIVE > QUIT LIVE**.
   - On a Windows-based system, choose **FILE > EXIT**.

   Ableton Live will automatically close.

That completes this exercise.

Chapter 3

Audio Recording Concepts

...*What You Need to Record Audio*...

This chapter introduces you to the fundamentals of audio recording. We begin with a discussion on the basics of sound, including the principles of frequency and amplitude. We then discuss various types of microphones and their uses, before diving into a discussion on multi-track recording. We follow this by examining the process of converting audio signals between the analog and digital domains and exploring the role of the audio interface in this exchange. The characteristics of analog and digital audio discussed in this chapter are important considerations when it comes to optimizing your results with any DAW.

Learning Targets for This Chapter

- Understand the audio characteristics of frequency and amplitude
- Understand characteristics of traditional microphones and USB microphones
- Understand the purpose of multi-track recording
- Understand the basic principles of analog-to-digital conversion
- Recognize the role of the audio interface in modern digital recording

Key topics from this chapter are illustrated in the Ableton Live Audio Production Basics Study Guide module available through the Elements|ED online learning platform. Sign up for a free account at ElementsED.com.

The process of recording audio has been commonplace for over 60 years, ever since magnetic tape-based recording processes began replacing earlier processes of inscribing a signal to a physical surface. In just the past 20 years or so, tape-based recording has itself largely given way to digital recording technology.

Regardless of the technology used to capture an audio recording, certain fundamental characteristics remain unchanged, including how an audio event moves through an acoustic environment and how that event is translated into an electrical signal that can be recorded. Recording in digital simply adds a dimension of collecting discrete measurements of the signal and storing those measurements as binary information.

The Basics of Audio

To understand the process of recording audio, it helps to first understand the basics of audio and sound waves. Sound waves are created when a physical object vibrates, causing a variation in the surrounding air pressure. By way of example, consider what happens when a guitar string is plucked.

The vibration of the guitar string, as it moves back and forth in a repeating cycle, causes a displacement of air particles. This results in a cyclical variation in air pressure, known as compression and rarefaction. The air pressure increases (compression) and decreases (rarefaction) in a way that corresponds directly to the pattern and frequency of the string's vibration. This cycling pattern of air pressure is referred to as a sound wave.

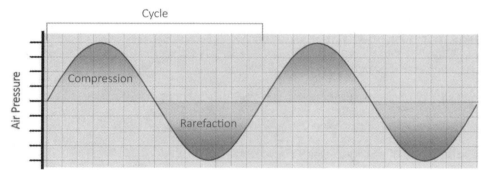

Figure 3.1 Cycles of compression (increasing air pressure) and rarefaction (decreasing air pressure) represented in a sound wave

Frequency

The frequency of the cycle of sound pressure variations determines our perception of the pitch of the sound. If the guitar string vibrates quickly, the air pressure variations will likewise cycle quickly between compression and rarefaction, creating a high-pitched sound. If the guitar string vibrates slowly, the air pressure will cycle slowly, creating a low-pitched sound.

We measure the frequency of these changes in *cycles per second* (CPS), also commonly denoted as *Hertz* (Hz). These two terms are synonymous—15,000 CPS is the same as 15,000 Hz. Multiples of 1,000 Hz are often denoted as kilohertz (kHz). Therefore, 15,000 Hz is also written as 15 kHz.

The open "A" string on a guitar vibrates at a frequency of 110 Hz in standard tuning. Playing the A note on the 12th fret produces vibrations at 220 Hz (one octave higher).

The range of human hearing is generally accepted to be between 20 and 20,000 cycles per second. This is also commonly denoted as 20 Hz to 20 kHz. Therefore, to capture a full-spectrum recording, we must be able to represent all frequencies within this range.

Amplitude

The intensity of a sound, or the amount of change in air pressure it produces, creates our perception of the loudness of the sound. A sound's intensity is represented by the amplitude (height) of the sound wave. We measure amplitude in *decibels* (dB).

The decibel scale is defined by the dynamic range of human hearing, from 0 dB at the threshold of hearing to approximately 120 dB at the threshold of pain. The decibel is a logarithmic unit that is used to describe a ratio of sound pressure. As such, the decibel does not have a linear relation to our perception of loudness. An increase of approximately 10 dB is required to produce a perceived doubling of loudness.

By way of example, the amplitude of ordinary conversation is around 60 dB. Increasing the amplitude to around 70 dB would essentially double the loudness, similar to what you might face trying to hold a conversation in a room with the TV or radio on. Increasing the amplitude to 80 dB would double the loudness again, such as you might experience trying to hold a conversation in a crowded room.

Microphones

The most common technique used to capture an acoustic audio event for recording is to place a microphone somewhere near the sound source. Microphones come in many shapes, sizes, and price points, so it can help to know a little bit about how they work before selecting the microphone or microphones you plan to use.

Traditional Microphones

All microphones are transducers, meaning they change energy from one form to another. A traditional analog microphone simply converts the variations in air pressure into variations in an electrical current. Three basic types of microphones are commonly available to make this conversion: dynamic mikes, ribbon mikes, and condenser mikes.

Dynamic Microphones

A dynamic microphone consists of a diaphragm attached to a coil of wire encircling a magnet. When sound waves hit the diaphragm, it vibrates back and forth, thereby moving the coil of wire back and forth over the magnet. This movement induces a current in the coil, with compression creating a positive voltage and rarefaction creating a negative voltage.

The advantages of dynamic microphones are that they are rugged and durable. They can also handle high input signals without distortion. Dynamic microphones are commonly used in both studio and live performance environments.

Ribbon Microphones

A ribbon microphone consists of a thin metal foil or other conductive ribbon material suspended between the positive and negative poles of a magnet. As sound waves hit the ribbon, they cause it to vibrate; the ribbon's movement within the magnetic field induces a current within the ribbon itself.

The advantages of ribbon microphones include a warm, smooth tone and an ability to capture high-frequency detail. Some ribbon microphones can be delicate and expensive. Ribbon mikes tend to be better suited for use in a studio environment rather than live performance.

Condenser Microphones

Condenser microphones consist of a thin conductive diaphragm (front plate) affixed a small distance in front of a metal back plate. The two plates are typically energized with a fixed charge. When sound waves hit the front plate, the diaphragm vibrates, varying the distance between the two plates. This varies the capacitance of the circuit and creates an opposite change in electrical voltage.

The capacitance is the ratio of the electric charge on the plates to the voltage difference between them. The capacitance increases as the plates move closer together and decreases as the plates move further apart.

Condenser microphones require power to operate, either from a battery or a phantom power source. Phantom power is typically 48 volts DC applied through a microphone cable to the condenser mike.

Mixers, microphone preamps, and audio interfaces commonly provide the option to enable phantom power, often denoted as +48V on the device.

The advantages of condenser microphones include a wide, smooth frequency response and sharp, detailed transient response. Condenser mikes are particularly well suited for acoustic instruments and cymbals. Although condenser mikes are very popular for in-studio recording, they are also commonly used in live sound production as well, particularly as drum overhead mikes.

USB Microphones

Another option to consider for home or studio use with your DAW is the USB microphone. USB mikes function the same as their traditional microphone counterparts, in that they convert acoustical energy into electrical energy. However, a USB mike goes a step further using two additional circuits: a built-in preamp and an analog–to–digital converter. This additional circuitry allows the microphone to be plugged directly into your computer for use with your DAW without requiring a separate audio interface.

For information on the analog-to-digital conversion process, see "Moving Audio from Analog to Digital" later in this chapter.

Most USB microphones are condenser mikes, although a few dynamic USB mikes are also available.

Popular choices among USB microphones include:

- **Blue Microphones**
    - **Snowflake USB Microphone**—Designed for mobile desktop recording at an affordable price ($)
    - **Snowball USB Microphone**—Designed for podcasting and other desktop recording ($$)
    - **Yeti USB Microphone**—Designed for studio vocals, musical instruments, voiceovers, field recordings, podcasting, and desktop recording ($$$)
    - **Spark Digital Lightning Microphone**—Designed for studio and desktop recording for voiceovers, vocals, and instruments at a premium price ($$$$)
- **CAD U37 USB Studio Condenser Recording Microphone**—Designed for desktop recording and voiceovers, vocals, and instrumental recording on a budget ($)
- **Audio-Technica ATR2100-USB Cardioid Dynamic USB/XLR Microphone**—Designed for stage and studio use at a reasonable price ($$)
- **Rode NT-USB USB Condenser Microphone**—Designed for studio-quality recording of vocals, instruments, and voiceovers with built-in monitoring control ($$$)
- **Apogee Mic 96k Professional Quality Microphone**—Designed for high-resolution studio-quality recording at a premium price ($$$$)
- **Rode Podcaster USB Dynamic Microphone**—Designed for studio-quality podcasting and other desktop recording at a premium price ($$$$)

Other Considerations

Aside from the characteristics of different types of microphones described above, a few other considerations should come into play when selecting the right mike for your needs.

Polar Pattern

A microphone's polar pattern describes how the microphone responds to sound coming from different directions. Typical choices include omnidirectional mikes, bidirectional mikes, and unidirectional mikes.

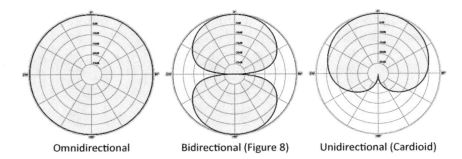

Figure 3.2 Common polar patterns: omnidirectional, bidirectional, and unidirectional

Omnidirectional. An omnidirectional mike will pick up sound equally from all directions. While this may be desirable for a microphone placed in the middle of a large conference table, it is usually not the best choice for studio recording where the goal is to isolate a sound source.

Bidirectional. Bidirectional mikes pick up sound primarily from the front and rear of the microphone, rejecting sound coming from the sides. This kind of mike may be desirable for across-the-desk interviews, two-part vocal recordings, or similar recordings of two opposite-facing sound sources. By rejecting sound from the sides, this type of mike will limit the amount of room acoustics and unwanted ambient noise in the recording.

Unidirectional (cardioid). Unidirectional mikes are the most popular choice for recording isolated sound sources. These are also commonly referred to as cardioid microphones. Unidirectional, or cardioid, polar patterns are designed to pick up sound from the front of the microphone, while rejecting (or attenuating) sound coming from the sides or rear of the microphone.

Proximity Effect

Most unidirectional and bidirectional microphones exhibit a noticeable bass boost when used within a few inches of a target sound source. This phenomenon is known as the proximity effect and often results in an unwanted boomy sound. You can hear the effect by speaking into a microphone as you move up close to the front of the grill.

If you are aware of a mike's proximity effect, you can adjust the mike placement or roll off the bass frequencies with an equalizer to compensate.

Basic Miking Techniques

Microphone placement and position have a large impact on how well a microphone will perform at capturing the desired signal. The overall goal when setting up a microphone is to target the sounds you want to record while rejecting the sounds you don't. Several accessories are available to help with this process.

Microphone Accessories

The microphone accessories discussed below can assist you with mike placement, noise reduction, and sound isolation.

- **Mike Stand**—USB microphones designed for desktop use often come with a small stand or clip. For miking in a studio or open environment, however, you will need to invest in a full-sized mike stand. You might also consider a stand with a boom extension or gooseneck add-on for better reach and flexibility. You'll find a variety of stands and extensions readily available from your local music store or online retailer.

- **Windscreens and Pop Filters**—Windscreens and pop filters are designed to reduce unwanted noise resulting from wind or bursts of air hitting the microphone. These problems are especially prevalent for on-location field recordings and close-miked recordings of vocal or dialog performances.

> ⓘ In vocal performances, bursts of air are commonly produced from words that begin with "p" or "b" sounds. These bursts, or plosives, can ruin a recording if they hit a microphone's diaphragm directly.

- **Shock Mounts**—Shock mounts are designed to suspend a microphone, isolating it from any mechanical vibrations that may affect the mike stand. Such vibrations are commonly caused by loud music, noise carried through a floor or stage, or nearby thumps and bumps.

- **Acoustic Shields and Vocal Booths**—To improve sound isolation, especially for vocal and dialog recordings, you can consider using an acoustic shield. These products are designed to reduce the influence of ambient noise, room acoustics, and reflected sound on a recording. Acoustic shields are most effective at attenuating mid– and high–frequency audio.

    A more extreme approach is to build or install a fully–enclosed vocal booth or isolation room. These will be more effective at reducing unwanted noise than a shield, but they can also be expensive and complex to design.

    Note, also, that both shields and booths may colorize the audio, boosting or cutting certain frequencies of the performance. You may be able to correct for the colorization using an equalizer; however, you will need to determine whether the isolation gains are worth the trade-off.

Where to Shop for Audio Gear

While locally-owned music stores are becoming scarce, audio gear has never been easier to shop for. In addition to ubiquitous Guitar Center storefronts, numerous online retail outlets are available. Online options include guitarcenter.com, sweetwater.com, musiciansfriend.com, and vintageking.com, among others. Almost any audio equipment can be found with a quick Internet search!

Basic Setup and Connections

Setting up your microphone or microphones is a simple matter of finding the right mike position and establishing a connection to your computer.

- **Microphone Position**—The location of your microphone during a recording can have a dramatic influence on the recorded signal. Generally speaking, the closer the mike is to the target sound source, the "cleaner" the recording will be. That is to say, the recording will include more of the desired sound and less ambient noise.

  However, as noted above, certain mikes exhibit a proximity effect when placed close to their sound source. This can colorize the sound in ways that may be unwanted, depending on the situation.

- **Microphone Connection**—The way that your microphone connects to your computer will vary, depending on the type of mike you are using. Most analog microphones use a cable with XLR connectors (3-pin) to attach to your audio interface. The audio interface will in turn connect to your computer using a USB cable or other common digital computer connection.

Figure 3.3 Microphone XLR connection to an audio interface

A digital USB microphone will connect directly to your computer, with no audio interface required.

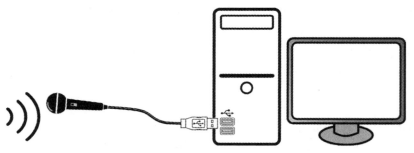

Figure 3.4 USB microphone connection to a desktop computer

Multi-Tracking and Signal Flow

The benefits of recording to a DAW are many. For starters, it allows you to record multiple takes and keep the best parts of each take. Additionally, it gives you the ability to edit and process a recording to improve the final results. But one of the greatest advantages is the freedom to record numerous isolated parts, each on its own track (known as multi-track recording).

What Is Multi-Track Recording?

The concept of multi-track recording has been around for many decades and is a fundamental aspect of audio production. Using this production method, each audio source is isolated as an individual recording on its own track. Each of these recorded tracks can then be edited and processed discretely. By playing back all of the individual tracks simultaneously, you can create a mix of the multiple sound sources as a final output.

By way of example, a music recording might include individual tracks for vocals, guitar, bass guitar, keyboards, and drums. By isolating each part on its own track, the volume levels, pan settings, EQ settings, and other processing can be set independently for each component.

Another advantage of multi-track recording is that individual parts can be recorded at different times. This is key if the performers cannot all be present in the same location at the same time. It also allows individual contributors to record multiple parts. For example, a musician can record both guitar and piano parts for a song. Or a voice actor can record dialog lines for multiple characters in a script for an animated short.

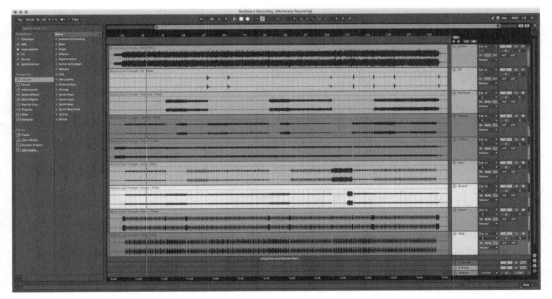

Figure 3.5 Multi-track music production session in Ableton Live

Not all recordings require multiple tracks. In some cases, you may want to record a mix of multiple sources to a single track. Some examples might include recording a live ensemble performance with a pair of microphones routed to a single stereo track or using an external mixer to sum multiple drum microphones to a stereo track. One reason for doing this is to simplify the recording and the number of tracks in your session. You might also use this approach if you do not have enough inputs on your system to record from many microphones at the same time.

Recording Signal Flow

Recording onto a computer involves routing a signal through various processing stages. This routing path is commonly referred to as the signal flow. Here, we will focus on a typical recording signal flow, from the sound source to the storage device (HDD, SSD, or cloud storage).

The sound originates at a source, such as a guitar, and travels through an environment by way of variations in air pressure, as discussed above. The air pressure changes are picked up by a microphone, converting the sound into an electrical signal. The electrical signal next travels down the microphone cable to an audio interface.

Within the audio interface, the signal is boosted, using a pre-amplifier (or *pre-amp*). The pre-amp is used to create a healthy signal level for recording. Next, the signal is sampled, using periodic measurements of the electrical voltage. The measured values are represented as strings of binary numbers and are passed along to the computer. At the computer, the binary numbers are stored on a hard disk drive or solid-state drive or cached and uploaded to cloud storage.

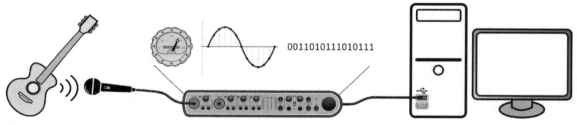

Figure 3.6 Recording signal flow from a guitar to a computer's hard drive

Moving Audio from Analog to Digital

As mentioned above, the practice of recording audio to a computer involves converting an electrical signal into individual, discrete binary measurements. This is the process known as analog-to-digital conversion.

Analog Versus Digital Audio

The variations in electrical voltage produced by a microphone represent a continuously changing analog audio waveform. In order to represent that information on a computer, the waveform must be converted into binary numbers that the computer can understand.

> ### What Is Binary Data?
> Binary data is comprised of binary digits, or bits. Each binary digit can represent only two possible values: *zero* or *one*. (At the most basic level, computers store and work with values using "switches" that are either off or on. A bit value of *zero* can be stored by toggling the associated switch off, while a bit value of *one* can be stored by toggling the switch on.)

> In order to store values larger than one, computers use strings of multiple binary digits. For example, using a string of four binary digits, a computer can store values from zero (0000) to fifteen (1111). Using a string of eight binary digits, a computer can store values from zero up to 255. Using a 16-bit string, the computer can store values up to 65,535.

Once an audio signal has been converted to digital, the waveform can be stored, read, and manipulated by the computer. The digital audio file is comprised of individual measured values (samples) stored as strings of binary digits.

The Analog-to-Digital Conversion Process

Converting an analog signal to digital involves two critical parameters: the sample rate and the bit depth.

Sample Rate. The sample rate refers to the frequency at which the incoming electrical signal is measured by the audio interface. The sample rate must be high enough to accurately represent the original analog signal.

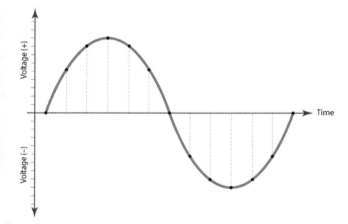

Figure 3.7 Sample intervals used to measure the changing voltage of an incoming audio signal

The sample rate required to accurately record a given signal is driven by a fundamental law of analog-to-digital conversion, known as the *Nyquist Theorem*. The Nyquist Theorem states that, in order to represent a given audio frequency, each cycle of the waveform must be sampled at least two times. Or, stated another way, the sample rate must be at least two times the frequency of the audio you wish to record.

Because we generally want to record full-spectrum audio – from 20 Hz to 20,000 Hz – the Nyquist Theorem tells us we need to use a sample rate of at least 40,000 Hz (or 40kHz): twice the upper end of this range. Most modern digital audio systems use a minimum sample rate of 44.1kHz.

Bit Depth. The bit depth (or word length) refers to the number of binary digits included in each measurement string. The more binary digits used, the more accurate the measurements will be. Low bit-depth recordings exhibit a loss of audio detail—the result of a reduced dynamic range.

> **What Is Dynamic Range?**
>
> Dynamic range is the difference between the quietest signal that a system can represent and the loudest signal that the system can represent. Dynamic range is measured in decibels (dB). This is a relative measurement, not an absolute loudness value. The maximum absolute loudness of a system depends on the amount of amplification applied at the output; however, the dynamic range remains constant at all loudness levels, since it measures the relative *difference* between two levels.

You can estimate the dynamic range of a system by multiplying the bit depth by six. Said another way, each binary digit that is added to the word length provides approximately 6 dB of dynamic range.

The useful range of loudness values for speech and music is generally considered to be from 40 dB on the quiet end to 105 dB on the loud end: a 65-dB dynamic range. In order to produce that dynamic range, a digital audio system must provide a minimum 11-bit word length (65 dB divided by 6 dB per bit). Additional dynamic range is required to allow for headroom before clipping and to accommodate any inherent noise floor in the system. Most modern digital audio systems use a minimum bit depth of 16 bits.

The Audio Interface

As mentioned previously, an audio interface is a device used to provide analog-to-digital conversion for the audio coming into a system. It also provides digital-to-analog conversion for the audio playing out of the system. The audio interface is what allows you to get sound into and out of your computer.

Most audio interfaces also provide one or more microphone pre-amps. Pre-amps generally include a gain control that allows you to adjust the incoming signal and optimize record levels.

Audio Interface Considerations

A multitude of audio interfaces are available today, providing an enormous array of options. To narrow down your choices, it helps to decide on the characteristics that are important for your recording needs.

Some factors to consider include:

- What kind of input and output (I/O) connectivity do you need?
- What level of sound quality are you looking for?
- What sort of budget are you working with?

I/O Connectivity

If you plan to work exclusively in Ableton Live Intro or Ableton Live Lite, you can consider an audio interface with up to 8 inputs, which is the maximum that Ableton Live Intro currently supports. If you plan to use Ableton Live Standard or Ableton Live Suite, you can consider one or more audio interfaces with a combined total of up to 256 inputs. For the greatest flexibility, look for an audio interface with microphone (XLR) inputs.

If you are operating on a limited budget, you can consider an audio interface that includes just 1 or 2 microphone inputs. These may be supplemented with additional line- or instrument-level inputs. Line/instrument inputs can be useful for recording from an electric guitar or keyboard connected directly to the audio interface along with your miked source(s).

Most audio interfaces will include balanced stereo outputs, via a pair of ¼-inch jacks, for connecting to studio monitors. Many also include a stereo headphone jack with volume control. A headphone output is useful for monitoring playback from your DAW while recording: monitoring through headphones instead of your studio speakers helps prevent the playback from bleeding into a live microphone.

Sound Quality

Generally speaking, the "quality" of a digital audio file is a factor of the sample rate and bit depth used for the recording. But the recorded results are also influenced by the quality of the microphone pre-amps and the analog-to-digital (or A/D) converters in the audio interface.

Other factors also have a significant impact, including the recording environment and the quality and type of microphone(s) being used to capture the audio.

All of that is to say that you should not consider the specifications of sample rate and bit depth in isolation when seeking to optimize the quality of your recordings.

> ### When to Use High-Resolution Audio
>
> For a simple recording project, you may find yourself using a utilitarian microphone to record audio that has a limited dynamic range (such as when recording dialog for a podcast or webinar). You may also be recording in an untreated room or office environment. In such cases, recording with a high sample rate and bit depth will not have a beneficial impact: recording at 44.1kHz and 16-bit will certainly be sufficient.

> Other times, you may be recording something with a wider dynamic range, such as a musical performance. This type of recording is often done using higher-quality microphones in a home-studio environment with at least some degree of acoustic treatment. In this case, you might want to consider using higher-resolution audio for better results and more editing flexibility. Recording at 24-bits is recommended in this situation to preserve the quality and dynamic range of the audio being captured.
>
> In cases where you find yourself working in a professional studio environment—with top-of-line microphones, boutique pre-amps, and other high-end gear—you will almost certainly want to record at a sample rate of 96kHz or above (and 24-bit) to capture all the subtle nuances of the performance and the character of the associated equipment.

 Recording at 32-bit floating point is typically only necessary when you plan to do extensive additional processing to the audio files after recording. You will hear no audible difference in the quality of the captured audio over recording at 24-bits.

Nearly all of the audio interfaces on the market today support sample rates up to 96kHz, with many supporting up to 192kHz. Likewise, 24-bit A/D conversion is a near universal standard for bit depth. This means that your search for an audio interface for Ableton Live will not need to be based on the supported sample rate and bit depth of the device, as just about any device will fit the bill.

 Ableton Live supports audio with sample rates up to 192kHz.

Instead, look for an audio interface with a construction quality and price point that complements your recording goals and budget. For higher-quality audio, look for devices from "brand name" companies that have a reputation for high-end gear…but expect to pay a premium for them.

Budget

The ultimate deciding factor often comes down to budget. But you can weigh the options, such as I/O connectivity and brand reputation, against your budget to find the right compromise.

For example, if you need only a single mike input, you may opt for a high-end audio interface to maximize the sound quality. On the other hand, if having 4 mike inputs is an absolute requirement, but sound quality is less of a concern, you might select a more utilitarian audio interface instead to stay within budget.

In the end, be sure to consider the budget for your entire recording setup as a whole. You will want to balance the cost of your audio interface against the other considerations covered in this chapter to ensure that you are not spending too much in one area and neglecting others.

Working Without an Audio Interface

In some cases, you may decide to work without an audio interface. You could do this as an interim measure until you can save up for the audio interface you want. You could also do this as way to run your DAW when you are away from your audio interface, such as when on the road with a laptop.

In either case, Ableton Live allows you to utilize the built-in audio on your computer (if available) in lieu of a dedicated audio interface.

On Mac-based systems, no special hardware or software setup is required to work without an audio interface. Launching Ableton Live without an interface connected will cause the software to use the computer's built-in audio inputs and outputs as the audio engine, automatically connecting the available inputs and outputs on your computer to the software.

On some Windows-based systems, Ableton Live will be able to utilize the computer's input and output options right out of the box. Other systems may need a driver installed to allow Ableton Live to access the computer's sound card options. The go-to resource for this is ASIO4ALL, a universal ASIO driver for Windows Driver Model (WDM) audio. ASIO4ALL is an independently developed freeware utility.

ASIO stands for Audio Stream Input/Output, a computer soundcard driver protocol developed by Steinberg Media Technologies.

In most cases, you simply need to download and install the ASIO4ALL software. Once installed, ASIO4ALL will allow you to run Ableton Live on a Windows machine with no audio interface connected and play back through the computer's built-in speakers or headphone jack.

To download ASIO4ALL, go to www.asio4all.com and download the latest version in the language of your choice.

Review/Discussion Questions

1. When a sound event occurs, what are the changes in air pressure called (i.e., the areas where air pressure increases and decreases)? (See "The Basics of Audio" beginning on page 66.)

2. What is meant by the frequency of an audio waveform? How does the frequency of a sound wave affect our perception of the sound? (See "Frequency" beginning on page 66).

3. What characteristic affects our perception of the loudness of a sound? How is loudness measured? (See "Amplitude" beginning on page 67.)

4. What is the purpose of a traditional microphone? What three basic types of microphones are commonly available? (See "Traditional Microphones" beginning on page 67.)

5. How are USB microphones different from traditional microphones? What is the advantage of using a USB microphone over a traditional microphone? (See "USB Microphones" beginning on page 68.)

6. What are the differences between omnidirectional microphones, bidirectional microphones, and unidirectional microphones? Which is the most common type of mike for recording isolated sound sources? (See "Polar Pattern" beginning on page 69.)

7. What are some accessories available to help you with mike placement, noise reduction, and sound isolation when recording? (See "Microphone Accessories" beginning on page 71.)

8. How is multi-track recording different from recording to a single audio file? What are some advantages of multi-track recording? (See "What Is Multi-Track Recording?" beginning on page 73.)

9. What two parameters are critical for the analog-to-digital conversion process? What minimum specifications are used for these parameters in modern digital audio systems? (See "The Analog-to-Digital Conversion Process" beginning on page 75.)

10. What is the purpose of an audio interface? (See "The Audio Interface" beginning on page 76.)

11. What are some factors to consider when selecting an audio interface for use with your DAW? (See "Audio Interface Considerations" beginning on page 76.)

12. How is working without an audio interface useful? What additional hardware or software options might you need in order to run Ableton Live without an interface? (See "Working Without an Audio Interface" beginning on page 79.)

 To review additional material from this chapter and prepare for certification, see the Ableton Live Audio Production Basics Study Guide module available through the Elements|ED online learning platform at ElementsED.com.

Exercise 3

Selecting Your Audio Production Gear

🎧 Activity

In this exercise, you will define your audio production needs and select components for a home studio based around Ableton Live software. By balancing your wants and needs against a defined budget, you will be able to determine which hardware and software options make sense for you.

🕓 Duration

This exercise should take approximately 10 minutes to complete.

◈ Goals/Targets

- Identify a budget for your home studio
- Explore microphone options
- Explore audio interface options
- Explore speakers/monitoring options
- Consider other expenses
- Identify appropriate components to complete your system

Getting Started

To get started, you will create a list of needs and define an overall budget for your home studio setup. This budget should be sufficient to cover all aspects of your initial needs for basic audio production work. At the same time, you'll want to be careful to keep your budget realistic so that you can afford the upfront investment. Keep in mind that you can add to your basic setup over time to increase your capabilities.

Use the table below to outline your basic requirements and to serve as a guide when you begin shopping for options. Place an **X** in the appropriate column for your expected needs in each row.

Do not include MIDI gear in this table, as we will address that separately.

 This exercise assumes that you own a compatible computer with built-in speakers for playback. Do not include a host computer in this table.

Function or Component	Not Required	Minimum Configuration	Maximum Configuration for Ableton Live Intro
Audio Interface: Output Channels (Playback)	–	1-2 Inputs	8 Inputs
Audio Interface: Input Channels (Recording)	–	2 Outputs (for stereo playback)	8 Outputs (for stereo playback and output to external gear)
Output Device(s)	Built-in Computer Speakers	Headphones	Stereo Monitor Speakers
Input Device(s) (Microphones)	–	USB Microphone	XLR Microphone(s) (specify type and number)
Accessories and Other (List or Describe) (Stands, Soundproofing, Software Add-Ons, etc.)			

Available Budget for the Above: _____

Identifying Prices

Your next step is to begin identifying prices for equipment that will meet your needs. Using the requirements you identified above as a guide, do some Internet research at an online music retailer of your choice to identify appropriate options for each of the items listed in the table below. You may also want to browse some manufacturers' websites for more information.

Component	Manufacturer and Model	Unit Cost
Audio Interface		
Microphones		
Headphones		
Monitor Speakers		
Accessories/Other		
	TOTAL	

Finishing Up

To finalize your purchase decisions, compare the total in the table above to the budget you allocated. If you find that your budget is not sufficient to cover the total cost, you will need to determine which purchase items you can postpone or consider bundle options. On the other hand, if you have money left over in your budget, you can consider upgrade options.

Bundle Options

One way to save some money is to look for gear bundles. These options can be much more affordable than buying the same components separately. Following are some good examples of bundles from Focusrite:

- Scarlett Solo Studio
- Scarlett 2i2 Studio

Each of the above bundles includes a Focusrite audio interface, a large–diaphragm condenser microphone, and a pair of closed-back studio headphones. As an added bonus, each bundle also includes a copy of Ableton Live Lite.

For more information, check out the details on the Focusrite website:

- Scarlett interfaces: https://us.focusrite.com/scarlett-range

Chapter 4

MIDI Recording Concepts

...What You Need to Record MIDI...

Understanding MIDI is essential to producing music using a DAW. In this chapter, we'll take a look at some basic MIDI concepts that will get you started creating tracks. We'll begin by taking a quick look at the history of MIDI and how the MIDI protocol works. Next we'll check out some of the types of MIDI controllers that are currently available. Then we'll look at how to set up the controller to communicate with your DAW. This will lead us into a discussion on the fundamental differences between MIDI and audio data. Finally, we'll take a quick look at how to begin working with virtual instruments.

⌖ Learning Targets for This Chapter

- Understand the basic history of MIDI
- Learn about the MIDI protocol
- Recognize types of MIDI controllers
- Learn how to set up your controller and DAW
- Recognize the differences between MIDI and audio
- Begin tracking with virtual instruments

Key topics from this chapter are illustrated in the Ableton Live Audio Production Basics Study Guide module available through the Elements|ED online learning platform. Sign up for a free account at ElementsED.com.

The term MIDI stands for Musical Instrument Digital Interface. MIDI is a protocol for connecting electronic musical instruments, computers, and other devices, allowing them to communicate with one another.

The MIDI standard was developed in the 1980s to allow musical performance information to be shared among and between devices. The standard provides specifications for describing musical events, such as note values and durations, allowing musical performances to be represented numerically.

A Brief History of MIDI

Before MIDI, there was really no standardized way to have synthesizers and other electronic musical instruments communicate with each other. Early voltage-controlled analog synthesizers by Bob Moog and Don Buchla (et al.) could send *voltage* information to each other. But they couldn't communicate more complex musical ideas, such as a discrete note with all of its pertinent information (pitch, amplitude, duration, etc.).

Figure 4.1 A Buchla 200e analog modular synthesizer

Digital Control

Another major drawback to analog synthesizers was that individual sounds could not be saved; favorite sounds had to be rebuilt by hand to be reused. Practitioners would actually make little drawings that showed where cables were patched and the position of each knob. (See Figure 4.2.) Some would even take Polaroid pictures so they'd have a photograph of the patch. (Interestingly, this way of working is making a huge comeback with the current Eurorack synthesizer craze!)

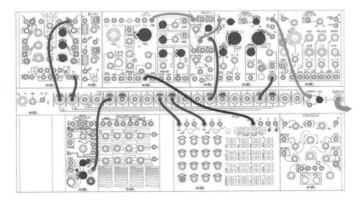

Figure 4.2 A patch "sheet" from a modern Eurorack analog synth

By the early 1980s, despite still using analog circuitry to generate sounds, most synthesizers had made the move to digital control for things like patch storage and recall (saving sounds and recalling them later). This created the possibility of having a common digital communication standard that would allow an electronic music instrumental to speak to any other instrument, regardless of the manufacturer.

The Birth of MIDI

In 1981, a proposal for a "Universal Synthesizer Interface" (USI) was presented at a meeting of the Audio Engineering Society (AES). This lead to the original MIDI 1.0 Specification. The specification was finalized in 1983 by a group that would become the MIDI Manufacturers' Association. The term *MIDI* was coined, as an acronym for Musical Instrument Digital Interface.

Image courtesy of Perfect Circuit Audio

Figure 4.3 Sequential Circuits Prophet 600 (1982)

At the 1983 National Association of Music Merchants (NAMM) trade show, the first public unveiling of MIDI occurred. A MIDI connection was demonstrated between a Sequential Circuits Prophet 600 (shown in Figure 4.3) and a Roland Jupiter 6. Soon, every major manufacturer had added MIDI connectivity to their synthesizers, drum machines, sequencers, and other devices.

The MIDI standard also helped pave the way for computer-based music. In 1984, Passport Designs released their MIDI/4 sequencer that ran on a personal computer.

The MIDI Protocol

The complete MIDI protocol is quite complicated, but it can be informative to take a look at the basics. MIDI messages can be divided into three basic categories: MIDI notes, program changes, and controller messages.

MIDI Notes

MIDI notes are probably the most important part of the protocol. Note data is the building block for almost all the work you'll do with MIDI.

One thing to understand about MIDI messages is that almost all parameters have a range of values spanning from 0 to 127. So, for instance, the pitch of a MIDI note is represented using a range of 0 to 127. A value of 0 represents the "C" note two octaves below the lowest note of a piano. We refer to this note as C-2.

Middle C is typically assigned a note value of 60 and referred to as C3 (although some manufacturers use C4 instead). Figure 4.4 provides a diagram of note values.

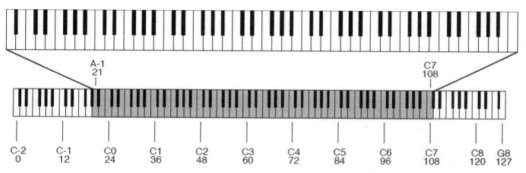

Figure 4.4 A standard piano keyboard with note names and MIDI note numbers

The other important bit of note information is the velocity, or how hard a note is played. Velocity is also represented using a range of 0-127, with the value of 64 falling right in the middle (sort of a *mezzo forte* in musical terms). Playing a note lighter will result in a lower velocity value, whereas playing harder will result in a higher value.

So, a MIDI note is a combination of a note number (pitch) and a velocity number. We refer to this combination as a **NOTE ON** command, because it activates the note in a technical sense. The Note On is eventually followed by a **NOTE OFF** command, which tells the device to stop playing the note. The Note Off command is the same note number, paired with a velocity of zero. The Note On and Note Off events are really all we need to send a note between two MIDI devices.

Program Changes

As mentioned above, patch storage and recall was an early advantage of digital control for synthesizers. The MIDI protocol improved upon this, enabling patch change commands (also known as program changes) to be sent to and from devices.

Like other MIDI messages, patch change commands use a range of values from 0 to 127. Thus, most banks of sounds on a MIDI keyboard will have 128 sounds. Table 4.1 shows an example bank of 128 sounds that many synthesizers support. This is part of the *General MIDI* standard.

Table 4.1 General MIDI Instrument Families

Program Change #	Family Name	Program Change #	Family Name
1 – 8	Piano	65 – 72	Reed
9 – 16	Chromatic Percussion	73 – 80	Pipe
17 – 24	Organ	81 – 88	Synth Lead
25 – 32	Guitar	89 – 96	Synth Pad
33 – 40	Bass	97 – 104	Synth Effects
41 – 48	Strings	105 – 112	Ethnic
49 – 56	Ensemble	113 – 120	Percussive
57 – 64	Brass	121 – 128	Sound Effects

Controller Messages

The final category of MIDI messages we'll consider is the category of controller messages—also known as MIDI continuous control (CC) messages. These are messages that can be used for purposes such as capturing the expressiveness of a MIDI performance (pitch bend or mod wheel movements) or performing studio mixing tasks (volume or pan changes). MIDI CC data represents an essential part of how keyboardists actually *play* their instrument.

Table 4.2 provides a list of standard MIDI continuous control defaults. Notice that certain controller numbers are "undefined," leaving room to make custom assignments.

Table 4.2 MIDI Controller Numbers (Excerpt)

Controller Number	Hex	Controller Name
0	00h	Bank Select
1	01h	Mod Wheel
2	02h	Breath Controller
3	03h	Undefined
4	04h	Foot Controller
5	05h	Portamento Time
6	06h	Data Entry MSB
7	07h	Main Volume
8	08h	Balance
9	09h	Undefined
10	0Ah	Pan

MIDI Controllers

To get started working with MIDI, it helps to have a MIDI controller. A MIDI controller is a device that generates MIDI performance data and transmits it to another device, such as a DAW, for processing. You can use a MIDI controller for recording MIDI notes and performance data into your DAW or for triggering a sound module or virtual instrument.

A variety of different MIDI controllers are available today. The most common designs continue to be based on the piano. With origins in the 18th century, the familiar black and white piano keyboard is a staple in the MIDI controller market. However, the last decade or so has seen an explosion in MIDI controller innovation. Newer concepts include drumpad controllers (inspired by the drum machine designs of the 1980s), grid controllers (originally introduced to augment the grid-based interface of Ableton Live), and mixer controllers (borrowing their faders and knobs from established analog and digital mixer designs).

Digital Pianos and Synthesizers

Many keyboard-based devices on the market can be classified as *tone-generating keyboards*. These are keyboards capable of generating their own sounds. Although generally designed for stand-alone performance, most tone-generating keyboards also provide MIDI connectivity. This allows you to use the keyboard to generate MIDI performance data and record it on your DAW. It also allows the keyboard to receive MIDI data from elsewhere and use it to trigger the device's onboard sounds.

Tone-generating keyboards generally fall into two categories: digital pianos and synthesizers:

- **Digital Pianos**—Digital pianos are exactly what the name implies: they look like a scaled-down version of a real piano. Digital pianos typically feature a full 88-key keyboard (often with simulated piano action known as *hammer action*). This type of keyboard is designed to compete head-to-head with real acoustic pianos; thus they are very simple to operate.

  Digital pianos generally offer a limited range of sounds (such as piano and organ). However, more expensive models may include additional sounds and auto-accompaniment features that can generate bass lines and drum patterns. Perhaps the most distinguishing feature of a digital piano is the presence of built-in speakers, which synthesizers rarely offer. (See below.)

  Figure 4.5 Yamaha Arius Digital Piano

- **Synthesizers**—The term "synthesizer" encompasses a wide range of vintage and modern instruments. The term was originally used to describe analog devices that employed a combination of oscillators and filters to create sounds. These devices generated approximations of acoustic instruments (such as flute or violin) as well as otherworldly sounds that didn't bear resemblance to traditional instruments.

  But synthesizers quickly evolved to encompass both analog and digital technologies, with newer incarnations capable of using older analog-style synthesis as well as modern digital techniques such as sampling, physical modeling, and frequency modulation (FM) synthesis.

  Modern synthesizers come in a variety of sizes, ranging from small, highly-portable two-octave (25-key) models to full 88-key piano–sized models.

Figure 4.6 Moog's classic MiniMoog synthesizer (1970)

If you already own a digital piano or synthesizer, look for a MIDI 5-pin connector or USB port for connecting it to your DAW. If you don't already own such a device, you can consider using a keyboard controller instead.

Keyboard Controllers

Keyboard controllers are essentially piano-style controllers that provide MIDI output, which you can use to create or record a MIDI performance. However, they do not include any onboard sound or tone-generating capabilities. This means that you cannot use a MIDI controller as a stand-alone instrument. Instead, the keyboard is designed to control other hardware synthesizers, sound modules, or virtual instruments.

The keyboard controller serves as a kind of universal MIDI input device. Using a piano keyboard as the performance interface, keyboard controllers enable any user with basic piano skills to create just about any kind of MIDI performance with little to no added learning curve.

Keyboard controllers come in a variety of sizes and price points. Users on a limited budget can consider a basic 25- to 37-key controller for under a hundred dollars. More sophisticated models with up to 88 keys can be found in the hundred-fifty to five hundred dollar range. These commonly include knobs, faders, and drumpads that can be mapped to control almost any function of the target instrument.

Figure 4.7 M-Audio Oxygen 49 keyboard controller

Drumpad Controllers

Most modern drumpad controllers are descended from Roger Linn's ingenious designs from the 1980s. The most iconic of Linn's designs was the Akai Professional MPC60, introduced in 1988. The 4×4 layout of pads that Linn pioneered has since become a standard pad arrangement for dozens of products.

The pads themselves are carefully designed to have a great feel that helps the player to perform dynamic drum and percussion grooves.

Figure 4.8 Akai Professional MPD218 drumpad controller

Grid Controllers

Today, the grid controller is probably the most popular controller type after the piano-style keyboard. Grid controllers have become popular for studio work, but they have also been embraced by live performers in a variety of genres, including EDM and hip-hop.

The grid controller design was inspired by Ableton Live's session view, with each pad used to trigger clips, play drum sounds, and adjust controls such as volume. Some grid controllers, such as Novation's Launchpad Pro and Ableton's Push 1 & 2, can be used to enter notes as well. These include options for restricting notes to a musical scale so that no wrong notes can be played! And, while grid controllers are generally optimized to work with Ableton Live, they can be used with an increasing number of DAWs.

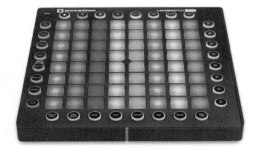

Figure 4.9 Novation Launchpad Pro grid controller

Alternate Controllers

You can also find some very innovative new controller styles on the market these days, offering alternatives to the standard keyboard, pad controller, or grid controller.

A primary focus of such alternative controllers currently is Multidimensional Polyphonic Expression (MPE). MPE is a new industry-wide standard that uses a separate MIDI channel for each touch. This channel-per-note configuration permits the controller to transmit discrete vibrato, glissando (pitch slides), note pressure, and other expressive information for each note that is played.

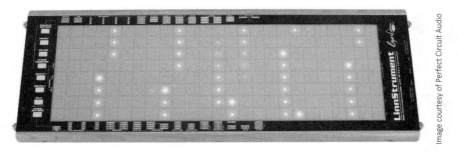

Figure 4.10 Roger Linn's Linnstrument is one example of an alternate controller

What to Look for in a MIDI Controller

Deciding which controller or controllers to purchase can be an overwhelming task! The most important thing is to determine which activities you will be performing frequently, which you'll be doing only occasionally, and which you won't be doing at all. Here are some suggestions based on the strength of various controller types for typical music production tasks:

- Entering Notes:
  - Keyboard (best choice)
  - Grid (good alternative)
- Playing Beats:
  - Drumpad (best choice)
  - Grid (good alternative)
  - Keyboard (also good)
- Arranging:
  - Grid

Purchasing a Keyboard Controller

If you decide to purchase a keyboard controller, you'll need to decide on a few other attributes. The most important is probably the number of keys (note range). Your needs here will generally be determined by your proficiency as a keyboard player.

If you are an accomplished keyboardist who plays with two hands simultaneously, you will probably feel most comfortable with a keyboard featuring 61 keys or more. A full-sized, 88-key controller will offer the widest note range and the best feel. On the other hand, if you play using just one hand at a time, you may be satisfied with a smaller 25-key model.

Aside from note range, you'll also want to consider a device's support for velocity sensitivity, aftertouch, pitch and mod wheel controls, and other mappable controls like knobs and sliders.

Velocity Sensitivity
As mentioned previously, representing how hard each note is played is often a key ingredient to the expressiveness of a performance. Most moderately sophisticated keyboard controllers will be velocity sensitive, meaning they measure how hard each key is struck. However, some budget models may not include this capability, representing a significant limitation for many types of MIDI recording.

Aftertouch
When available, aftertouch lets a performer add expressive inflections to a sustained note or chord. Aftertouch can be used to add vibrato, pitch bends, and swells by varying the pressure on the held keys. When using a virtual instrument that supports it, aftertouch can be essential to making the instrument sound realistic. (Think of guitar solos and horn parts, for example: how often are sustained notes completely static?) Here again, budget keyboard controllers often do not include support for this parameter. Be prepared to do some comparison-shopping when trying to get the most bang for your buck.

Pitch and Mod Wheel Controls
These are additional options for adding expressiveness to a performance. Pitch bend and mod wheel controls are common on synthesizers but may not be included on some MIDI controllers. A traditional pitch wheel adds pitch bend (portamento) control, allowing the player to bend a note (or chord) up or down in a continuously variable manner. Pitch bend controls are sometimes included in non-wheel format, such as in a joystick, knob, or ribbon format.

Modulation wheels (or mod wheels for short) are typically used to set a vibrato amount. However, the mod wheel can also be mapped to other parameters, such as volume swells, tremolo, and filter sweeps.

Other Mappable Controls
Additional buttons, knobs, and sliders may be included on mid-sized and full-sized keyboard controllers. These controls can be mapped to plug-in parameters and other MIDI controls. In some cases, they can even operate functions in your DAW, such as transport controls and volume faders.

Setup and Signal Flow

Once you've selected a MIDI controller, you'll need to establish communication between the controller and your computer. Almost all modern MIDI controllers use USB rather than an actual MIDI cable to communicate. Configuring such a device can be as simple as plug-and-play, if the device is USB class-compliant. However, some devices require that you install specific drivers built for your computer and operating system.

Plug-and-Play Setup

For "class-compliant" USB devices, no drivers are required to communicate with your computer. You can verify that a device is class-compliant by checking the specifications on the manufacturer's website. Alternatively, you can connect the device to your computer and launch the appropriate software utility to verify communication.

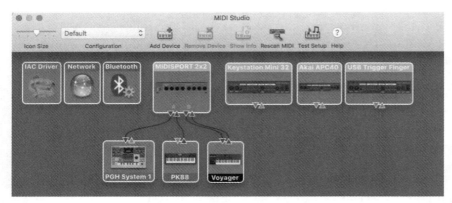

Figure 4.11 An example MIDI setup in Apple's Audio MIDI Setup utility; class-compliant devices appear here automatically

MIDI Cables and Jacks

If you are using a device that has regular MIDI connections (not USB), you will have to deal with MIDI cables and jacks. A MIDI cable is technically called a 5-pin DIN cable and has connectors as shown in Figure 4.12.

Figure 4.12 A pair of MIDI cables

MIDI data is actually transmitted using only two pins on the 5-pin connector.

An important aspect of a MIDI cable is that it can carry *16 channels* of information. That means you could have up to 16 patches playing on one synthesizer, and they could all be addressed individually using a single MIDI cable.

The jacks that MIDI cables plug into are typically called *ports*. These can include an input, an output, and sometimes a "thru" connection.

Figure 4.13 A typical configuration of MIDI ports on a device

The MIDI In port receives incoming MIDI data from an outside source; this is commonly used to play back a MIDI performance coming from a DAW or other source using the onboard sounds of the device. The MIDI Out port sends MIDI data out from the device, for recording to a DAW or for triggering a different sound module. The MIDI Thru port passes the signal from the In port directly out to another device, without modifying it by any performance being done on the device. This can be very handy if you have a small MIDI interface with only a couple of ports, or if you have a complex live performance setup and don't want to lug an interface around.

Using a MIDI interface

Although most modern controllers use USB, you may encounter a hardware device (such as an older synthesizer) that uses traditional 5-pin MIDI cables to communicate. If so, you'll need a way to connect the MIDI cables to your computer, because the computer will not have the appropriate jacks for MIDI. You can accommodate these connections using an audio interface with MIDI jacks (quite common on all but the smallest interfaces).

Figure 4.14 MIDI jacks on the back of a Focusrite 4i4 audio interface

As an alternative to an audio interface, you can use a dedicated MIDI interface to route MIDI data to your computer. MIDI interfaces come in a variety of shapes and sizes and can have as many as eight jacks for MIDI input and output.

Figure 4.15 M-Audio Uno MIDI interface

Considerations for Using Multiple MIDI Devices

Even the smallest studio will often have more than one MIDI device. This could entail any combination of keyboard, drumpad, grid, and mixer controllers. Aside from the ergonomic issues of placing multiple controllers in the physical space, you'll also need to consider whether you want the devices to control specific instruments or to all control the currently selected instrument.

In most DAWs, the ideal solution is to leave inputs on your MIDI tracks assigned to "All Ins" so that the currently selected instrument will respond to any controller that you touch. This makes it very easy to play keyboard for a piano part, then switch over to the drumpads to perform a beat, all without having to stop and change the MIDI routing.

MIDI Versus Audio

One of the more difficult concepts to grasp when you begin working with a DAW is the difference between MIDI and audio. While most novice audio producers have a decent understanding of how audio data is recorded and represented in a DAW, many do not have a similar understanding with regard to MIDI data.

What Is MIDI?

As discussed earlier in this chapter, MIDI is a protocol that facilitates communication between a huge range of hardware and software devices. MIDI data is not audio, but rather a series of messages that can communicate information such as notes (pitch), duration, velocity (how hard a key or pad is pressed), dynamics (volume), and more. A tone-generating hardware or software instrument can receive MIDI messages and convert them into audio data.

In other words, the MIDI data has the potential to become music that we can hear, but there's no way to listen to the MIDI data by itself.

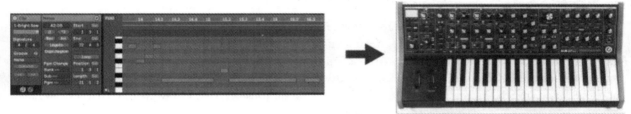

Figure 4.16 MIDI data (in piano roll format) must be sent to a tone-generating device to become music that we can hear

MIDI messages share some features with traditional music notation. Both contain the information necessary to play a piece of music, but neither is capable of actually becoming music on its own. Like MIDI, the notes of a musical score aren't audio, but instead represent the potential for audio once they are played using an instrument.

MIDI and notation contain information that can convey essential musical elements such as pitch, duration, velocity, and dynamics. Almost every DAW allows MIDI data to be viewed and edited as music notation (in addition to piano roll, event list, and other more modern formats).

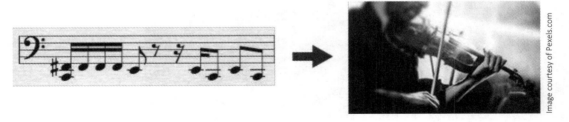

Figure 4.17 Music notation must be read by a musician to become music we can hear.

Monitoring with Onboard Sound Versus Virtual Instruments

If you are using a tone-generating keyboard such as a digital piano or synthesizer, you will need to decide whether you want to monitor from the onboard sound of the device or, alternatively, silence the onboard sounds and instead listen to a virtual instrument inside of your DAW.

It is rare that you will want to hear the sound of the controller while recording MIDI data into your DAW. (In fact, this can result in an annoying doubling of the controller audio with the DAW audio.) In the majority of situations, you'll want to monitor from the associated virtual instrument's output alone. This can be accomplished by turning off "Local Control" on the keyboard device. This setting will disable the internal sounds of the device while still sending MIDI data out to the DAW.

You may need to refer to the user guide or manufacturer's website to find the local control setting on your particular device.

Tracking with Virtual Instruments

Most modern music production studios incorporate software instruments, or virtual instruments, into their toolkit. In fact, a number of software manufacturers are known primarily for their virtual instrument offerings, including Native Instruments, Arturia, and Spectrasonics. Virtual instruments encompass a huge range of musical devices, including digital synthesizers, analog-modeling synthesizers, samplers, drum machines, and much more. They have become so ubiquitous in the DAW world that every major manufacturer offers a range of free or paid virtual instruments.

Figure 4.18 Native Instruments Kontakt sampler

Creating Tracks for Virtual Instruments

Working with virtual instruments in your DAW can seem a bit confusing at first. It's not quite as simple as just creating a track and pressing record! The first step to working with virtual instruments is to create the appropriate track type, which is usually referred to as an "MIDI" or "Instrument" track depending on the DAW. These tracks are capable of recording and editing MIDI data, routing the MIDI data to a virtual instrument, and then mixing the virtual instrument's resulting audio output.

Let's take a look at a couple of example configurations.

Virtual Instruments in Ableton Live

Routing MIDI to a virtual instrument in Ableton Live is quite similar to the process in many other DAWs, including Pro Tools, Cubase, and Logic.

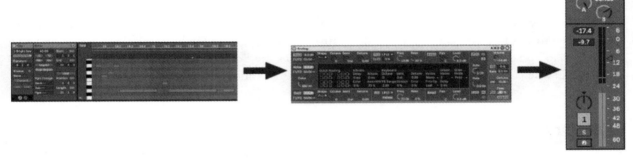

Figure 4.19 When MIDI is routed to the Analog device, the device's audio is sent to the track's Fader section.

In Figure 4.19, the MIDI track sends its MIDI data to the Analog device, which is inserted on the same track. Then, Analog converts the MIDI data to audio, with the sound being determined by the patch/preset that is currently selected. Next, the audio output from Analog is routed through the audio section of the MIDI track, where its volume, pan, and other audio attributes can be adjusted as desired. Finally, the audio output of the track is routed to the audio interface, so that it can be monitored through speakers or headphones.

Virtual Instruments in GarageBand

GarageBand provides another example of this concept, although the approach used in the software is somewhat different. (See Figure 4.20.) Note that GarageBand doesn't offer a mixer view, so the track is only visible in a horizontal layout that is similar to the Arrangement view in Ableton Live.

In GarageBand, the sound module is assigned by selecting the desired source from the Sound Library, rather than by placing a virtual instrument plug-in on the track.

Figure 4.20 Using a drum instrument in GarageBand

In Figure 4.20, we can see the MIDI data in the familiar piano roll format on the track. That MIDI data is being routed to the East Bay drum kit (whose controls are displayed at the bottom). The kit converts the MIDI data to audio. The audio output actually appears on the horizontal controls just to the left of the MIDI data, where the volume and pan attributes can be modified. From there, the audio goes to the interface output.

Summary

As you can see, the function of a virtual instrument is quite similar regardless of the DAW you choose. Professional-level DAWs offer sophisticated routing and additional views for working with MIDI data (as well as audio). By contrast, entry-level DAW options (such as GarageBand) aim to keep things as simple as possible. Regardless of where you begin, once you have gained basic familiarity with your chosen DAW, you'll find it relatively easy to apply that knowledge to any other DAW you should encounter.

Review/Discussion Questions

1. What were some of the drawbacks of historical analog synthesizers? (See "A Brief History of MIDI" starting on page 88.)

2. How did digital control of synthesizers pave the way for the development of MIDI? (See "A Brief History of MIDI" starting on page 88.)

3. What are the three basic categories of MIDI messages? (See "The MIDI Protocol" starting on page 90.)

4. What is the standard range of values for all MIDI messages? (See "The MIDI Protocol" starting on page 90.)

5. What MIDI note number is typically used to represent middle C on a piano? Is this number the same with every device? (See "The MIDI Protocol" starting on page 90.)

6. What are the two components of a MIDI Note On command? (See "The MIDI Protocol" starting on page 90.)

7. What is the name for a patch change command when using MIDI? (See "The MIDI Protocol" starting on page 90.)

8. What type of MIDI data can be used to increase the expressiveness of a MIDI performance? (See "The MIDI Protocol" starting on page 90.)

9. What are two categories of tone-generating keyboards? (See "MIDI Controllers" starting on page 92.)

10. Aside from keyboards, what are some other types of MIDI controllers? (See "MIDI Controllers" starting on page 92.)

11. What does it mean when a USB device is labeled "class-compliant?" (See "Setup and Signal Flow" starting on page 98.)

12. How can MIDI notes be converted to an audio signal that we can hear? (See "MIDI Versus Audio" starting on page 100.)

13. How can a virtual instrument device or plug-in be assigned in Ableton Live? (See "Tracking with Virtual Instruments" starting on page 102.)

 To review additional material from this chapter and prepare for certification, see the Ableton Live Audio Production Basics Study Guide module available through the Elements|ED online learning platform at ElementsED.com.

Exercise 4

Selecting Your MIDI Production Gear

∩ Activity

In this exercise, you will define your MIDI production needs and select components to complement the Ableton Live system you configured in Exercise 3. By identifying the type of MIDI production work you will be doing, you will be able to select an appropriate MIDI controller (or controllers) to meet your needs.

⏲ Duration

This exercise should take approximately 10 minutes to complete.

✛ Goals/Targets

- Identify a budget for your MIDI hardware
- Explore controller options
- Consider other expenses
- Identify appropriate components to complete your system

Getting Started

To get started, you will create a list of requirements for your MIDI controller and define an overall budget for your MIDI setup. Once again, make sure your budget is sufficient to cover all your immediate needs, while remaining realistic about what you can afford. You can always add to your basic MIDI setup over time with more devices and virtual instruments.

Use the table below to identify your basic MIDI requirements and to serve as a guide when you begin shopping for a controller. Place an **X** in the appropriate column(s) for each row.

Criteria	Rating			
	Novice / Hunt and Peck	Competent Beginner	Intermediate	Advanced
Piano Keyboard Skills				
Intended Uses	Drum Programming, Effects, Short Musical Parts, Loops, EDM	Background Music Production, Simple Keyboard Parts and Other Instruments	Lead Instrument Parts, 2-Handed Performances, Complex Arrangements	Orchestral Scores and Arrangements
	Basic MIDI Input	Velocity Sensitivity	Pitch Bend and Mod Wheel	Aftertouch
Feature Requirements				
Accessories and Other (List or Describe) (Keyboard Stands, Foot Pedals, etc.)				

If your **X**'s fall predominantly in the left half of this table, you may want to consider a drum pad or grid controller, or look for a budget-model keyboard controller. If your **X**'s fall predominantly in the right half, you should probably target a full-sized and fully featured keyboard controller. If your MIDI needs are extensive, you may want to consider selecting more than one type of controller to purchase.

Available Budget for MIDI Gear: _____

Identifying Prices

Your next step is to begin identifying prices for the type of controller and accessories that will meet your needs. Using the requirements you identified above as a guide, conduct some research online to identify available options. List potential matches in the table below, along with short descriptions (including number of keys, if applicable) and prices.

Type of Controller	Manufacturer and Model/Description	Price
Accessories/Other		

Finishing Up

To finalize your purchase decisions, total the cost of each option plus your required accessories and compare them to the budget you allocated. If you find that your budget is not sufficient to cover your preferred choice, consider a different option with the plan to upgrade in the future.

If you are able to purchase your first choice items and have money left over in your budget, you can look at software add-ons to supplement your collection of virtual instruments.

Chapter 5

Ableton Live Concepts, Part 1

...What You Need to Know to Get Started with Ableton Live...

This chapter introduces you to some basic operations and functions you need to be familiar with to get started working in Ableton Live. We cover how to access and use the main views and sections in Ableton Live and how to create tracks. We also cover basic navigation and selection techniques, including transport controls, zooming and scrolling operations, and uses for selections. These foundational concepts will help you get up and running and remain productive as you work.

⊕ Learning Targets for This Chapter

- Recognize uses for the Arrangement View and Session View
- Become familiar with the basic controls in each main view
- Learn how to create tracks and understand the supported track types in Ableton Live
- Learn how to navigate in the Arrangement view by scrolling and zooming
- Learn playback and selection techniques

 Key topics from this chapter are illustrated in the Ableton Live Audio Production Basics Study Guide module available through the Elements|ED online learning platform. Sign up for a free account at ElementsED.com.

Ableton Live Views and Sections

Ableton Live software provides two primary views that you can toggle between in the main window: the Arrangement View and the Session View. You can also optionally display each of these views in separate windows.

Within the main Live window are a number of other available views and sections that are used to accomplish various tasks. At the top of the window, you will find the Control Bar, which includes the transport controls for Ableton Live. Beneath this is the area where the primary views for Ableton Live can be displayed: the Arrangement View, the Session View, the Browser View, and the Detail View. The Detail View itself features two separate view options: the Device View and the Clip View.

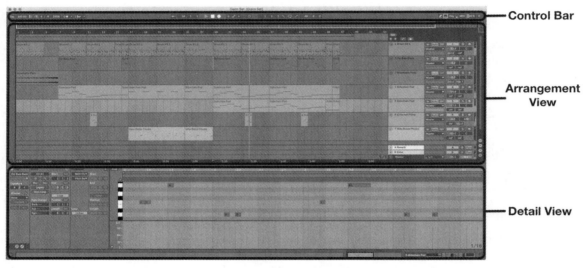

Figure 5.1 The main window displaying the Control Bar, Arrangement View, and Detail View (showing Clip View)

Control Bar

Ableton Live provides a variety of controls in the Control Bar area at the top of the main window. These controls cover a wide range of settings and displays for your Live set, including tempo, time signature, quantization settings, edit settings, keyboard and MIDI mapping, and performance indicators.

Figure 5.2 The Control Bar in Ableton Live

Among the most important features for getting started with Ableton Live are the transport and looping controls in the Control Bar. We discuss the primary transport functions later in this chapter, in the section on Basic Navigation.

Figure 5.3 The transport controls (Play, Stop, and Record) in the Control Bar

Arrangement View

The Arrangement View provides a horizontal timeline display of audio, MIDI data, video, and mixer automation for recording, editing, and arranging tracks. It displays waveforms for the audio clips, and MIDI note data for the MIDI clips. Each Audio and MIDI track in the Arrangement View features some of the same track sections as in the Session View (see below), including the In/Out section and Mixer section.

In/Out Section

The right portion of each channel strip in the Arrangement View provides controls for routing signals for the track. Depending on the type of track, these controls may include Audio Input and Output selectors, MIDI Input and Output selectors, and Monitor Mode controls. The Audio Input and Output selectors are used to route signals from your audio interface for recording or playback. The MIDI Input and Output selectors are used to route MIDI data from MIDI controllers into and out of Ableton Live. MIDI data can route either to internal instrument plug-ins or to external MIDI sound modules.

The Monitor controls for each track determine the sound source you will hear for that track. You can monitor either a live input connected to the track or playback of the existing clips on the track. The Monitor controls can generally be left on the **Auto** setting: this will automatically switch from monitoring the live input when a track is record armed, to monitoring clip playback when the track is not record armed.

Figure 5.4 The In/Out section of the Arrangement View

Mixer Section

The Mixer section of the Arrangement View provides a number of controls for mixing tracks. Each track displayed in the Mixer section has controls for panning, volume, and sends. The Mixer section also provides buttons for enabling record, activating/deactivating tracks, and toggling the track solo on and off.

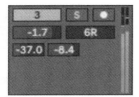

Figure 5.5 The Mixer section of the Session View

 The Volume control in the Mixer Section does not affect the input gain (record level) of a signal being recorded. The signal level must be set appropriately at the source or adjusted using a preamp or gain-equipped audio interface.

Session View

The Session View is the view that sets Ableton Live apart from most other audio software. When the first version of Ableton Live was released back in 2001, it was the Session View that caught the attention of music producers around the world. There was nothing else like it!

Figure 5.6 The Session View beneath the Control Bar (also shown: Browser View, and Detail View [showing Device View])

The Session View is designed to allow playback of clips and scenes asynchronously. This means you can play musical segments in any order and for any duration. This behavior is ideal for experimentation, arranging on the fly, and supporting live performances that may involve spontaneous changes.

> **Understanding Session View**
>
> In the Session View, clips can be placed in a track's clip slots and triggered to play or stop independently. Each clip on a given track plays through the devices inserted on that track, so that the same instrument and/or signal processing is used by all clips on the track. You can also trigger an entire row of clips across all tracks simultaneously by launching a Scene.
>
> Because Session View features non-sequential playback, it can be useful for sketching out songs and performing them into the Arrangement View. This is a classic Ableton Live studio production workflow. The Session View is also a powerful live performance tool. It can be used to play backing tracks for a live band, to perform DJ sets that are free-flowing yet tightly synchronized, or to fire off music cues and sound effects for a theatrical performance.

The Session View provides a number of controls grouped into sections. Several of these sections, such as the Clips Slots section and the Scenes section, are unique to the Session View. Other sections, such as the In/Out, Sends, and Mixer sections, can also be found in the Arrangement View.

 The Session View is discussed in detail in Chapter 7.

 You can toggle between the Session View and the Arrangement View in the main window by pressing the TAB key.

Additional Session View Sections
Several other Session View sections are available, including the Delay Section and the X-Fade Section. These functions are beyond the scope of this book.

Browser View
The Browser View can be displayed on the left-hand side of the main window. It can be accessed when either the Session View or Arrangement View is active.

The Browser view is used to locate clips, samples, instruments, audio and MIDI effects, plug-ins, and Max for Live devices. The Browser view is also used to access and install Ableton Packs (in Ableton Live 10 or later), to browse through user settings and current project files, and to quickly access folders located anywhere on your computer.

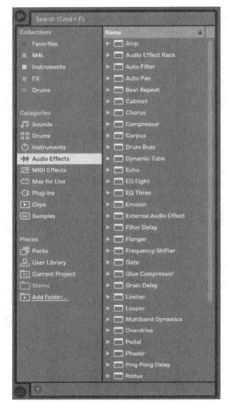

Figure 5.7 The Browser View in Ableton Live

Detail View

The Detail view can be displayed at the bottom of the main window. Like the Browser, the Detail View is accessible regardless of whether the Session View or Arrangement view is currently active. The Detail View is always operating in one of two modes: Device View or Clip View.

 To toggle the Detail View between Device View and Clip View, press Shift+Tab.

Device View

When the Detail View is set to Device View, it will display any devices inserted onto the currently selected track. These devices include instruments, audio effects and MIDI effects, as well as Max for Live devices.

Figure 5.8 The Device View

Clip View

When the Detail View is set to Clip View, it will display the contents of a currently selected clip. For audio clips, the Detail View displays the clip's waveform in the Sample Editor; for MIDI clips, it displays the clip's MIDI data in the MIDI Note Editor. The Clip View also features several property boxes that can be used to modify various aspects of the clip.

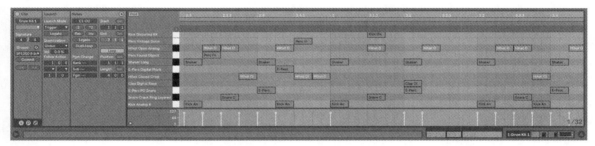

Figure 5.9 The Clip View in Ableton Live

Info View

The Info View is an optional display that is available as part of the Detail View. When shown, the Info View appears at the far left side of the Detail View.

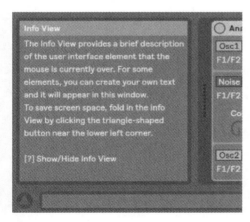

Figure 5.10 The Info View on the left side of the Detail View

When you position your mouse cursor above an element of the Ableton Live user interface, the Detail View will provide a brief description of that element in the Info View.

Showing and Hiding Views and Sections

Ableton Live lets you customize the display of the windows to accommodate your needs at any given point in your project.

 Either the Session View or Arrangement View must be visible at all times.

Showing/Hiding Views: The secondary Ableton Live views (Browser View, Detail View, and Info View) can be shown or hidden in one of two ways.

To show or hide views, do one of the following:

- Go to the **VIEW** menu and select a view to toggle it between shown or hidden.
- Click the triangular **SHOW/HIDE** buttons located in various places in the main window.

Figure 5.11 The triangular Show/Hide button for the Browser View

Showing/Hiding Sections: Both Session View and Arrangement View sections (such as the In/Out Section and the Mixer Section) can be shown or hidden in one of two ways.

To show or hide sections, do one of the following:

- Go to the **VIEW** menu and select a section to toggle it between shown or hidden.
- Click the **SHOW/HIDE** section buttons located in the bottom right corner of the Session View and Arrangement view.

Figure 5.12 The Show/Hide button for the sections in the Session View

Resizing Views: Several of the Ableton Live views can be resized to accommodate a particular task. These include the Browser View and Detail View.

To adjust the width or height of a view, follow these steps:

1. Position the mouse cursor over the view separator; the cursor will change into a double-headed arrow.
2. Click and drag on the view separator to adjust its position as needed.

Figure 5.13 Clicking on a view separator to resize a view

Overview (Arrangement View only)

The Overview section appears just below the Control Bar in the Arrangement View. It displays a miniature representation of all of the clips that are currently on tracks in the Arrangement view. You can click in the Overview to quickly navigate to a new position in the Set. You can also click and drag to change the horizontal or vertical zoom level.

Figure 5.14 The Overview section in Arrangement View

Rulers and Scrub Area (Arrangement View only)

Rulers are horizontal strips that appear in the Arrangement View, just above and below the Track Display area. Rulers provide measurement indicators to help you identify specific locations in your Set's timeline.

Figure 5.15 The Beat Time Ruler and Scrub Area in the Arrangement View

- **Beat Time Ruler.** This Ruler appears above the Track Display. It displays bar numbers when zoomed out but will show beats and sub-beats as you zoom in.

- **Scrub Area.** The Scrub Area sits just below the Beat Time Ruler and performs a number of functions. You can click in the Scrub Area to start playback from a specific location. You can also set loop start and end points by dragging the left or right side of the loop brace (or by entering values into the Loop Start and Loop Length fields in the Control Bar). In addition, you can right-click in the Scrub Area to insert a time signature change or to add a Locator.

 The loop brace and looping behavior are discussed later in this chapter.

- **Time Ruler.** The Time Ruler is a secondary ruler that appears at the bottom of the Arrangement View. You can configure this ruler to show Time in minutes:seconds or film and video frame rates including 24 frames per second (fps), 25 fps, and 29.97 fps (film, PAL, and NTSC, respectively). This can be helpful when working with imported video clips.

Working with Tracks

Once you've created a new Live Set, you will have a project with two MIDI tracks and two audio tracks by default.

 You can change this default behavior by modifying the default Live Set template in Live > Preferences (Mac) or Options > Preferences (Windows) under the File/Folder tab.

Tracks are where your audio and MIDI performances are recorded and edited. Audio and MIDI data can be copied and duplicated in different locations to create repeating patterns, to arrange song sections, or to assemble material from multiple takes.

Adding Tracks

To add additional audio or MIDI tracks to your session, choose **CREATE > INSERT AUDIO TRACK** or **CREATE > INSERT MIDI TRACK** from the menu bar.

Ableton Live supports the following five track types:

- **Audio tracks**—This track type lets you record or import audio clips and work with them in a waveform-based display format. Live supports mono (single-channel) or stereo (two-channel) audio.

- **MIDI tracks**—This track type lets you record, create, import, and edit MIDI clips. It can also host a virtual instrument plug-in.

- **Group tracks**—This track type functions as a container for other audio and MIDI tracks. It is ideal for submixing similar tracks together.

- **Return tracks**—Return tracks do not contain any audio or MIDI clips, but rather host audio effects. They are unique in that one Return track can process audio sent to it from any number of source tracks within the Live Set.

- **Master track**—The Master track controls the overall level of the audio output. It is included by default in all Live Sets. It can host audio effects and contains a master fader that controls the level of Ableton Live's final output.

Any combination of track types can be added to your Live Set, with the exception of the Master track, which is limited to one and is included in all Live Sets by default.

> ### Shortcuts for Creating Tracks
>
> Shortcuts exist to speed up the process of inserting new MIDI, Audio, Return, and Group tracks.
>
> - Use **COMMAND+T** (Mac) or **CTRL+T** (Windows) to insert a new Audio track.
> - Use **COMMAND+SHIFT+T** (Mac) or **CTRL+SHIFT+T** (Windows) to insert a new MIDI track.
> - Use **COMMAND+OPTION+T** (Mac) or **CTRL+ALT+T** (Windows) to insert a new Return track.
> - With one or more Audio or MIDI tracks selected, use **COMMAND+G** (Mac) or **CTRL+G** (Windows) to create a new Group track.

Clips Versus Files

Each recording you complete on a track in Ableton Live is stored as a separate audio file on disk. When you edit a clip, such as by trimming off the start or end of the recording, you create a subset of the audio for that clip. This smaller clip is a pointer that references a portion of the audio within the larger parent audio file.

By working with clips rather than the original files, Ableton Live can perform non-destructive audio editing. This means the original audio files are unaltered by the edits you perform. As a result, you can easily recover or reference "missing" audio after an edit by resizing a clip, applying crossfades, and using other techniques.

Basic Navigation

As a Set grows, both in overall length and in track number, it becomes increasingly important to be able to navigate through the project quickly. In this section, we cover basic navigation techniques for use with playback, recording, and editing in a project.

Playback in Arrangement View

When viewing a Live Set in the Arrangement View, the main Track Display window shows a horizontal timeline of the clips in your project. Playback starts from the location of the Arrangement Insert Marker (or playhead). This is a cursor indicated by a solid, blinking line that runs through one or more tracks. The playhead location is also indicated by an arrow in both the Beat Time Ruler and the Time Ruler. A numeric representation of the playhead's location is displayed in the Arrangement Position counter at the top of the Live Set.

Figure 5.16 The Control Bar, including the Arrangement Position counter (left), as well as the Play and Stop buttons

Click the **PLAY** button (or press the **SPACEBAR**) to begin playback based on the location of the playhead.

Playing Selections

You can also click and drag within the Track Display window to select a period of time. The selection is indicated by a highlighted zone that can span one or more tracks, with markers on each end in the Beat Time Ruler and Time Ruler. By making a selection on a track, you define an area for Ableton Live to perform a desired editing task, such as deleting, duplicating, or muting, among others.

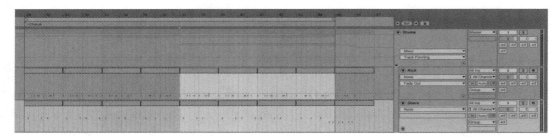

Figure 5.17 Selection from Bar 37 to Bar 45

Click the **PLAY** button (or press the **SPACEBAR**) to begin playback based on the start of the selection. Playback will continue beyond the end of the selection until you click the **STOP** button (or press the **SPACEBAR** a second time).

 Press COMMAND+SPACEBAR (Mac) or CTRL+SPACEBAR (Windows) to play *only* to the end of a selection.

Be aware that you cannot make a selection by clicking and dragging in the top part of a clip, as that will move the clip. Click and drag across empty space or within the bottom part of a clip to avoid moving it.

Alternatively, you can click on the top part of an Audio or MIDI clip, which selects the clip. With the clip selected, playback will begin from the start of the clip.

Cycling Playback with the Arrangement Loop

Playback in the Arrangement View is typically linear, proceeding from left to right. However, you can activate the Loop switch to have playback cycle repeatedly within a designated area.

The loop area is defined by the Arrangement loop brace—a horizontal strip in the Scrub Area with triangle markers at either end.

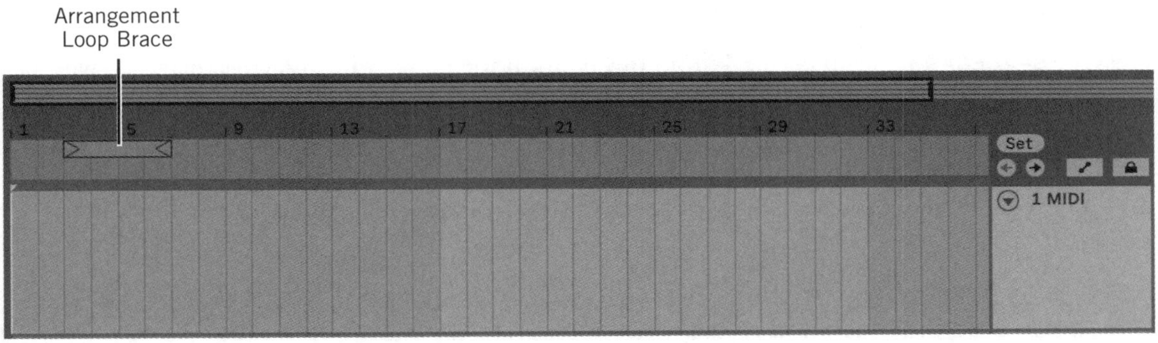

![Arrangement Loop Brace figure]

Figure 5.18 Arrangement loop brace in a new Set (shown inactive)

You can adjust the start and end points of the loop brace by dragging the corresponding triangle markers (start and end arrows) to the left or right. This lets you resize the area that will cycle during playback.

> (i) Cycling (or looping) during playback occurs only when the Loop switch is active.

To activate the Loop switch, do one of the following:

- Click on the **LOOP** switch in the Control Bar so that it becomes highlighted.

Figure 5.19 The Loop switch enabled in the Control Bar

- Choose **EDIT > ACTIVATE LOOP** or press **COMMAND+L** (Mac) or **CTRL+L** (Windows).

When the Loop switch is active, the Arrangement loop brace will become highlighted in the Scrub Area.

Starting and Stopping Playback

As mentioned above, playback will start from the Insert Marker location.

To play back a portion of your Live Set, follow these steps:

1. Click anywhere within an existing track to set the playback start point. (Avoid clicking on the top part of a clip, as that will select the clip.)

2. Press the **SPACEBAR** to begin playback from this point. If the Loop switch is active, playback will loop repeatedly whenever playback starts before or within the loop area.

3. To stop playback, press the **SPACEBAR** again.

4. To move to a different playback point in the track, click a new position and press the **SPACEBAR** again.

 You can stop playback and return your position to the start of the Live Set by clicking the Control Bar's Stop button twice.

Tracking the Playhead

At times, the playhead might move off screen. For example, if the **FOLLOW** scrolling option is not enabled (see "Using the Follow Scrolling Mode," below), the cursor will move off screen after reaching the edge of the Track Display window in the Arrangement View.

Live displays the cursor as a single vertical line in the Clip Overview at the top of the window in Arrangement View.

Using Follow Mode for Scrolling

Live includes a scrolling option called Follow mode that causes the playhead to always remain visible in the Track Display window. As the Live Set plays, the display scrolls to follow the playhead, meaning the playhead will never go off-screen.

To enable Follow mode, do one of the following:

- Choose **OPTIONS > FOLLOW**.
- Press **COMMAND+SHIFT+F** (Mac) or **CTRL+SHIFT+F** (Windows).
- Click the **FOLLOW** button in the Control Bar.

Figure 5.20 Clicking the Follow button in the Control Bar (shown active)

Follow behavior can be set to scroll continuously, keeping the playhead centered on screen, or to scroll one page at a time, scrolling each time the playhead reaches the right edge of the screen.

To change scrolling behavior:

1. Choose **LIVE > PREFERENCES** (Mac) or **OPTIONS > PREFERENCES** (Windows) and select the **LOOK/FEEL** tab.

2. Click on the button next to **FOLLOW BEHAVIOR** to toggle between the Page and Scroll options.

Zooming and Scrolling in the Arrangement View

You can use Ableton Live's zoom features to zoom into and out of a particular area within the set. Zooming in is often helpful when you need to examine a clip or waveform closely.

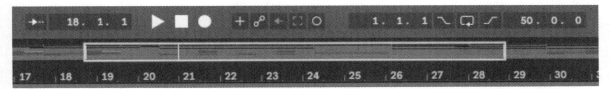

Figure 5.21 The beat-time ruler (bottom) and Overview area (with white highlight) in the Arrangement View

When working in the Arrangement View, you have several different ways to zoom and scroll. You can use the beat-time ruler, the Arrangement Overview, or keyboard shortcuts.

Zooming and Scrolling with the Beat-Time Ruler

The most common technique for zooming and scrolling in the Arrangement View is to use the beat-time ruler. When you position your mouse cursor over the beat-time ruler, a magnifying glass cursor will display.

To zoom in and out using the beat-time ruler:

- Click and drag vertically up and down to change the zoom level.

To scroll the display using the beat-time ruler:

- Click and drag horizontally to scroll left or right.

Zooming and Scrolling with the Arrangement Overview

The Arrangement Overview is like a "bird's eye" view of your entire set. It always shows the complete song in a miniature form. The current zoom level is indicated by a box outline in the Arrangement Overview (black or white, depending on your chosen color theme).

To change the zoom level using the Arrangement Overview, do one of the following:

- Click in the Arrangement Overview and drag vertically to zoom in or out.

- Position your cursor on a border of the Arrangement Overview rectangle; drag left or right with the bracket icon to modify the currently visible area.

To scroll using the Arrangement Overview, do one of the following:

- Click and drag within the Arrangement Overview rectangle to continuously scroll left or right.
- Click a different location in the Arrangement Overview to jump to that location in the Set.

Zooming and Scrolling with Keyboard Shortcuts

Experienced Ableton Live users generally prefer to use keyboard shortcuts for zooming and scrolling in the Arrangement View.

When zooming, each click of the shortcut keys zooms all tracks in by one level, with the Arrangement View centered on the insert marker.

To zoom using keyboard shortcuts:

- Press the **+** (plus) or **-** (minus) keys on the computer keyboard to zoom in and out, respectively.

 Press Z to fill the screen with the current Arrangement Time Selection. Press X to return to the previous zoom setting. Press X a second time to zoom out further.

To scroll left/right using keyboard shortcuts:

- Press **COMMAND+PAGE UP/PAGE DOWN** (Mac) or **CTRL+PAGE UP/PAGE DOWN** (Windows) to scroll left and right, respectively.

Optimizing Arrangement Height and Width

Ableton Live 10 introduced the ability to quickly optimize the height and width of the Arrangement View.

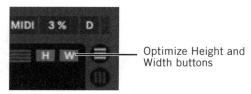

Optimize Height and Width buttons

Figure 5.22 The Optimize Arrangement Height (H) and Width (W) buttons

To optimize track height to fit the Arrangement View, do one of the following:

- Click the **OPTIMIZE ARRANGEMENT HEIGHT** button (**H**) in the upper right corner of the Arrangement View.

- Click the **H** key on the computer keyboard.

To optimize the song width to fit the Arrangement View, do one of the following:

- Click the **OPTIMIZE ARRANGEMENT WIDTH** button (**W**) in the upper right corner of the Arrangement View.
- Click the **W** key on the computer keyboard.

Review/Discussion Questions

1. Which Main View displays the audio waveforms and MIDI performances in a timeline display on tracks in your Ableton Live Set? (See "Ableton Live Views and Sections" beginning on page 112.)

2. How are tracks displayed in the Arrangement View? What are some track controls available in this window? (See "Arrangement View" beginning on page 113.)

3. What are some uses for the Browser View? (See "Browser View" beginning on page 115.)

4. What two main display modes are available as part of the Detail View? What is the difference between them? (See "Detail View" beginning on page 116.)

5. What are some of the controls that are available in the Control Bar? (See "Control Bar" beginning on page 112.)

6. What track types are supported in Ableton Live? (See "Working with Tracks" beginning on page 120.)

7. What track type can be used to host a virtual instrument? (See "Adding Tracks" beginning on page 120.)

8. What track type can process audio sent to it from any number of source tracks within the Live Set? (See "Adding Tracks" beginning on page 120.)

9. How many Master Tracks can be created in an Ableton Live Set? When is a Master Track created for a Set? (See "Adding Tracks" beginning on page 120.)

10. What are clips in Ableton Live? How is an audio clip different from an audio file? (See "Clips Versus Files" beginning on page 121.)

11. What are some ways to zoom in and out in the Arrangement View? What are some ways to scroll the Arrangement View to the left or right? (See "Zooming and Scrolling in the Arrangement View" beginning on page 125.)

12. What zooming and scrolling keyboard shortcuts are available in the Arrangement View? (See "Zooming and Scrolling with Keyboard Shortcuts" beginning on page 126.)

 To review additional material from this chapter and prepare for certification, see the Ableton Live Audio Production Basics Study Guide module available through the Elements|ED online learning platform at ElementsED.com.

Exercise 5

Configuring and Working on a Project

🎧 Activity

In this exercise, you will configure the display for the project you started in Exercise 2. You will also practice making selections and using the Zoom functions in Ableton Live. By zooming in, you will be able to make very accurate selections on track playlists and on the timeline, allowing you to perform edit operations with greater precision.

🕐 Duration

This exercise should take approximately 10 minutes to complete.

◈ Goals/Targets

- Adjust the width of the Browser View
- Hide the Help View and Detail View
- Optimize the Arrangement View Height and Width
- Make selections and practice zooming in the Arrangement View

---

Exercise Media

This exercise uses media files from the song, "Overboard," provided courtesy of Sacramento-area band The Pinder Brothers, with additional files provided by Eric Kuehnl from a remix of the song.

Written by: Matt Pinder; Performed by: The Pinder Brothers;
Produced by: Scott Reams and The Pinder Brothers*; Remix by: Eric Kuehnl and Frank D. Cook

The media provided for this course may be used for educational purposes only. No rights are granted to use the media for any other personal, commercial, or non-commercial purposes.

* The mix, processing, and media files have been adapted for use in the exercises contained herein.

Getting Started

To get started, you will reopen the multi-track project you worked with in Exercise 2. This will serve as the starting point for this exercise.

Open your existing Overboard project:

1. Launch Ableton Live, if it isn't already running. Once startup completes, the Default Set will be visible.

2. Select **FILE > OPEN LIVE SET**.

3. Navigate to the location where you saved your work in Exercise 2 and select your **Overboard01-xxx** project.

4. Click **OPEN**. The Set will open as it was when last saved.

Configuring the View

Before starting work on a Set, it can be helpful to adjust the views in Ableton Live to optimize the window displays. In this section of the exercise, you will adjust the width of the Browser View, and hide the Detail View for an unobstructed display.

Adjust the width of the Browser View:

1. Position the mouse cursor along the right-edge boundary of the Browser View, so that the double-headed cursor appears.

Figure 5.23 Double-headed cursor displayed at the border of the Browser View

2. Click and drag to the left to decrease the width of the Browser View. Keeping the Browser narrow will maximize space for your tracks.

Try to find a comfortable setting that still lets you read the names of items in the Browser when you select a category such as Instruments or Audio Effects.

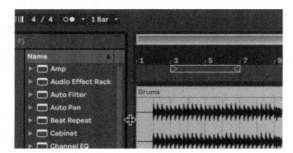

Figure 5.24 Browser View minimized to show items without using excessive space

Hide the Help View and Detail View:

1. Choose **VIEW > HELP VIEW** as needed to disable this option and hide the Help View. The panel on the right side of the window will collapse, providing more horizontal space for your track display.

2. Choose **VIEW > SHOW/HIDE DETAIL VIEW** to deselect this option and hide the Detail View. The panel at the bottom of the window will collapse, providing more vertical space for your track display.

Optimize the Height and Width of the Arrangement View:

- Click both the Optimize Arrangement Height (**H**) and the Optimize Arrangement Width (**W**) buttons in the upper right corner of the Arrangement View. (See Figure 5.25.) The tracks will resize and zoom as needed to fit within the available screen space. (See Figure 5.26.)

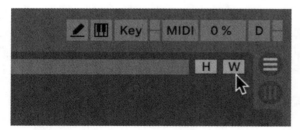

Figure 5.25 Clicking the H and W buttons in the Arrangement View (buttons shown enabled)

Figure 5.26 Arrangement View after configuring the display settings

Making Selections, Zooming, and Editing

The tracks in this Set represent a segment of a remix of the song "Overboard" by The Pinder Brothers. To get familiar with the audio in this project, you will start by selecting audio ranges to set the playback location and duration.

Select just the guitar solo on the LeadGtr track:

- Click in the lower half of the **LeadGtr** track around Bar 17, and drag across the active waveform area to the end of the song. Do not include the "silent" part at the end of the audio clip.

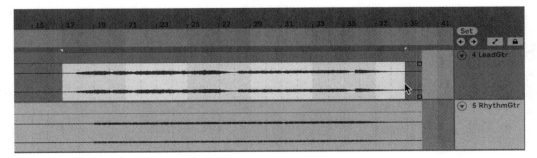

Figure 5.27 Selection on the LeadGtr track

Adjust the zoom for a better view:

- Press the **Z** key to zoom to the Arrangement Time Selection.

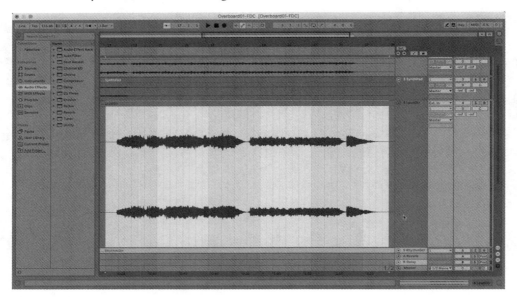

Figure 5.28 The Arrangement Time Selection zoomed to fit in the window

> ⓘ If the Z key does not zoom in on the selection, you may have the computer MIDI keyboard enabled. Press M to disable this function and try again.

Play the Set to hear the guitar solo:

- Press **SPACEBAR** or click the **PLAY** button in the Control Bar. Playback will begin at the start of the selection. Allow playback to continue through the end of the song.

The lead guitar part clashes a bit with the synthesizer part that plays during the second half of this section. To resolve this, you will remove the conflicting part of the **LeadGtr** track.

Select and remove audio on the LeadGtr track:

1. Click and drag in the lower half of the **LeadGtr** track, beginning at Bar 28 and continuing through the first half of Bar 35 (**Start:** 28.1.1; **End:** 35.3.1). This will select the portion of the track that conflicts with the synth lead. (See Figure 5.29.)

> ⓘ Use the Beat-Time Ruler at the top of the Arrangement View as a visual indication of the Bar.Beat locations where your selection should begin and end.

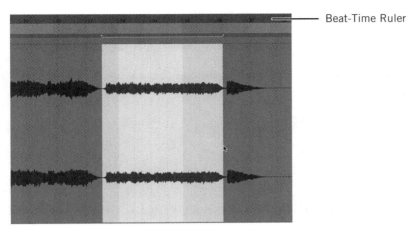

Figure 5.29 Selection on the LeadGtr track corresponding with the area where the synth lead should be prominent

2. Press the **DELETE** or **BACKSPACE** key on your computer keyboard to remove the selected audio.

Zoom back out to see the whole Arrangement:

- Press the **X** key to return to the previous zoom setting, with all tracks displayed on screen.

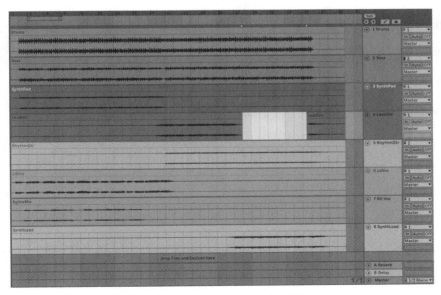

Figure 5.30 Tracks display after zooming out with the X key

 You may need to press the X key multiple times to zoom back to the original zoom setting.

Play through the solo sections of the arrangement again to hear the results:

1. Click on the **LeadGtr** track at Bar 17 to position the Insert Marker before the guitar solo.

2. Press the spacebar to begin playback and listen through to the end of the song. Note that the SynthLead part is now unobscured.

3. Press the spacebar again when finished to stop playback.

Finishing Up

Feel free to experiment further with selections and playback. Practice making selections and playing different parts of the song. When finished, return to the start of the project and save the work you've done.

Finish your work:

1. With the transport stopped, press the **STOP** button in the transport section of the Control Bar again to position the Insert Marker at the beginning of the project.

2. Choose **FILE > SAVE LIVE SET AS** to save a copy of the Set.

3. In the resulting dialog box, name your finished Set **Overboard02-xxx**, where xxx is your initials.

4. Click **SAVE** to complete the operation.

5. Exit Ableton Live by choosing **LIVE > QUIT LIVE** (Mac) or **FILE > EXIT** (Windows).

    Ableton Live will automatically close.

That completes this exercise.

Chapter 6

Ableton Live Concepts, Part 2

...*What You Need to Know to Work with Ableton Live*...

We begin this chapter with an overview of recording audio and MIDI performances using Ableton Live. Next, we explore options for importing existing audio and MIDI data into your session. And finally, we take a look at basic editing techniques that can be used to fine-tune audio and MIDI performances.

◎ Learning Targets for This Chapter

- Learn how to record audio
- Learn basic MIDI recording techniques
- Learn how to import audio and MIDI
- Learn basic techniques for working with clips and selections

Key topics from this chapter are illustrated in the Ableton Live Audio Production Basics Study Guide module available through the Elements|ED online learning platform. Sign up for a free account at ElementsED.com.

When starting a new Ableton Live Set, the first step typically involves getting some audio or MIDI data into your Set. This can be accomplished in a number of ways, but generally you'll either record new clips directly onto an Audio or MIDI track in your project or import existing clips from various locations on your computer.

Setting Up for Recording

When beginning a music project, it can be beneficial to set the project tempo for Ableton Live prior to starting to record. This will allow you to use the Metronome as a tempo reference during recording, helping the musicians keep the performance consistently in time. Additionally, recording with the Metronome enabled will ensure that the music aligns to the Beat-Time Ruler. As a result, any selections and edits you make can easily be aligned to the music by working with Snap to Grid enabled.

 The project tempo defaults to 120 beats per minute (BPM) in Ableton Live.

To specify the project tempo, do the following:

1. Click in the Tempo display within the Control Bar. The default value of **120.00** will become highlighted.

   Figure 6.1 Setting the project tempo

2. Do one of the following to change the tempo:
   - Type in the tempo value that you'd like to use (in BPM).
   - Click and drag up or down using the mouse.
   - Press the Up/Down Arrow keys.

With the tempo set correctly, you can enable the metronome to provide a rhythmic click according to the project tempo.

To enable the Metronome:

- Click the **Metronome** button in the Control Bar.

  Figure 6.2 Setting the project tempo

Recording Audio in Arrangement View

Recording audio is one of the most fundamental aspects of working with any modern DAW. Working in Ableton Live gives you non-linear access to all of the audio in your project. This means you can jump directly to any location without needing to fast–forward or rewind like you would with a tape recorder.

Ableton Live provides a truly intuitive workflow for quickly getting your musical ideas recorded onto a track. And thanks to the huge storage drives included with most modern computers, you'll rarely have to worry about running out of recording space.

Designating a Punch-In Point

Before you begin recording you'll need to specify where you want the record pass to begin. Audio engineers call the beginning of the record pass the "punch-in" point.

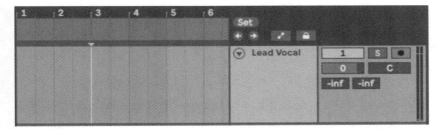

Figure 6.3 The insert marker will function as a punch-in point with no specified stopping point.

To specify a punch-in point:

- Single-click on a track to position the insert marker. During a record pass, Ableton Live will begin recording at this location and will continue recording until you manually stop the transport.

Record-Enabling Tracks

Once you know where you want to record, you'll need to decide which tracks you want to record on. A common beginner mistake is to assume that Ableton Live will record on the track where you've made a selection or positioned the insert marker.

In reality, the position of the insert marker determines only the time location where you will record, but it has no bearing on which track (or tracks) will be affected. To specify a destination track you simply need to record enable ("Arm") the desired track in the Arrangement View or Session View.

To record-enable a track:

- Click the **ARM** button in the track's Mixer section. (See Figure 6.4.)

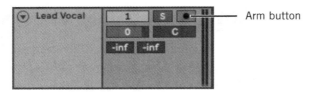

Figure 6.4 A record-enabled track in the Arrangement View

Monitoring Record-Enabled Audio Tracks

When a track is not record-enabled, you can typically press **PLAY** and hear the clips that are positioned on that track. However, listening to (or *monitoring*) record-enabled tracks is a bit more complicated. Ableton Live's default Monitor setting is Auto. With this setting selected, a record-enabled track will prioritize the playback of existing clips during playback, and you will not hear the live input from your audio interface. This can be a little confusing, especially if you are trying to audition a part before pressing record.

Fortunately, you can exclusively monitor the live input by setting the track's Monitor setting to "In."

To set a track's Monitor setting to In:

1. Locate the desired track.
2. Click the **IN** button in the track's In/Out section. Both the In button and Track Activator switch will turn blue.

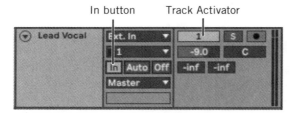

Figure 6.5 A track with monitoring set to In

 The Track Activator is a button displaying the track number, located to the left of the track's Solo (S) button in the Arrangement View. This switch will Mute a track when deactivated.

The standard Ableton Live workflow is to leave the Monitor setting on Auto. When you begin recording with Auto-monitoring enabled, Ableton Live will automatically switch over to monitoring the live input and will ignore the existing clips on the track.

> **Auto-monitoring**
>
> Using the Auto-monitoring setting lets you use pre-roll to start playback earlier than the desired punch-in point. This will allow you to hear playback from the track before the take begins, which is generally a more musician-friendly approach. It also allows you to listen back to the last recorded take without switching monitoring modes or disarming the track.

Initiating a Record Take

Once you have record-enabled one or more tracks, you have several options to initiate a record take in Ableton Live.

To initiate a record take:

1. Do one of the following to arm the Set for recording:

    - Click the **ARRANGEMENT RECORD** button in the Control Bar.
    - Press the **F9** key.

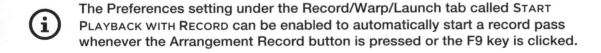

Figure 6.6 The Transport controls in the Control Bar

2. Do one of the following to begin recording:

    - Click the **PLAY** button in the Control Bar.
    - Press the **SPACEBAR**.

> (i) The Preferences setting under the Record/Warp/Launch tab called START PLAYBACK WITH RECORD can be enabled to automatically start a record pass whenever the Arrangement Record button is pressed or the F9 key is clicked.

Stopping a Record Take

When you initiate a record take with only a punch-in point specified, the record take will continue until you manually stop it. There are several techniques for stopping a record take in Ableton Live.

To immediately stop a record take, do one of the following:

- Click the **STOP** button in the Control Bar.
- Press the **SPACEBAR**.

To stop recording without stopping playback:

- Click the **ARRANGEMENT RECORD** button in the Control Bar.
- Press the **F9** key.

Often a musician will make a mistake and you will want to delete the clip(s) from the most recent record pass. To do this in Ableton Live, use the Undo command. Obviously, you'll want to use this option only if you're absolutely certain that the take is not worth keeping.

To use Undo to delete clips from most recent record pass, do one of the following:

- Press **COMMAND+Z** (Mac) or **CTRL+Z** (Windows).
- Choose **EDIT > UNDO RECORD**.

Recording MIDI in Arrangement View

Although the MIDI communication protocol is more than 30 years old, the importance of recording and editing MIDI data is more relevant today than ever. Advances in virtual instrument design and computer processing speed over the past decade have made virtual instruments an indispensable part of the modern music production workflow.

The fundamental workflow for recording MIDI is similar to that for recording audio. However, some important differences apply in MIDI production. But don't worry; you don't need a PhD in synthesis to start making music with the MIDI feature set in Ableton Live!

Monitoring a MIDI Controller

The monitoring workflow is one area that differs significantly when recording MIDI as compared to audio. This is because an audio signal can be monitored directly, whereas MIDI data must be routed to an instrument to become an audible signal. The routing can seem a little complicated at first, but once you understand the basics you'll be able to take advantage of this powerful technology.

 For basic information on routing MIDI data to virtual instruments, see "Tracking with Virtual Instruments" in Chapter 4.

Routing MIDI data to a virtual instrument is slightly different than routing audio signals. On a MIDI track, the I/O input is used to select the MIDI input device, whereas the I/O output is used to route the resulting audio.

Figure 6.7 The I/O view on a MIDI track

 Ableton Live automatically assigns the MIDI output to the first virtual instrument plug-in assigned on a MIDI track.

Using MIDI Arrangement Overdub

MIDI Arrangement Overdub is a powerful option for recording MIDI data. When disabled, each successive record pass results in a new MIDI clip on the track. In this mode, recording MIDI data is essentially the same as recording audio: you designate a punch-in point (such as by positioning the insert marker), record arm the target track, and click the **ARRANGEMENT RECORD** button (or press **F9**) to begin recording. If you are not satisfied with a completed record pass, you can use the Undo command to discard the recorded clip or simply record a new pass over the top to replace it.

With MIDI Arrangement Overdub enabled, however, successive record passes add MIDI data, merging it into the existing clip. This mode is great for recording the left hand and right hand parts of a piano performance separately, for example, and combining them both in the same clip.

MIDI Arrangement Overdub is also an essential tool for recording drum parts. You can start by recording just one or two drum pieces in the first pass (such as kick and snare) and then merge in additional pieces on subsequent passes (such as the hi-hat). (See Figures 6.8 and 6.9 below.)

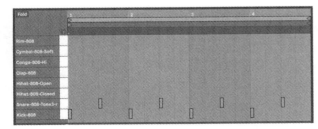

Figure 6.8 The first pass of a drum recording with just kick and snare

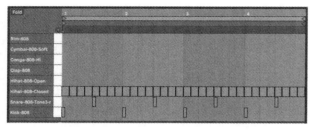

Figure 6.9 The second pass of a drum recording with hi-hat overdubbed into the clip

To enable MIDI Arrangement Overdub:

- Click the **MIDI ARRANGEMENT OVERDUB** button (plus sign) in the Control Bar.

Figure 6.10 The MIDI Arrangement Overdub button in the Control Bar (highlighted yellow)

Importing Audio and MIDI

If you're not ready to record in Ableton Live, or if you'd like to use existing media from your hard drive instead, you can import audio or MIDI files into your Ableton Live project to get started. There are a number of ways to import audio and MIDI files into Ableton Live.

Supported Audio Files

Before you attempt to import an audio file, you'll want to make sure that it is in a format that can be imported into Ableton Live.

The following file types can be imported:

- AIFF
- AIFC
- WAV
- SDII
- MP3
- AAC
- OGG Vorbis
- FLAC

Importing from the Desktop

An easy way to get started importing is to use the drag–and–drop method. Ableton Live fully supports drag–and–drop importing of audio and MIDI files from the Mac Finder or Windows Explorer. Simply select the desired files on your computer and drag them into your Ableton Live project. (We look at the different destinations available for dropped files below.)

Importing from the Browser

Another way to import audio and MIDI files into Ableton Live is to use the built-in Browser. The Browser can function similarly to a Mac Finder or Windows File Explorer. Like the Finder and File Explorer, you use the Browser to browse and search any mounted volume on your computer. However, the Browser offers features that are tailored to working with audio and MIDI files.

To display the Browser view, do one of the following:

- Select **VIEW > SHOW BROWSER**.

- Press **OPTION+COMMAND+B** (Mac) or **ALT+CTRL+B** (Windows).

- Click on the **SHOW/HIDE BROWSER** button in the upper left corner of the main window.

    The Browser view will open.

Figure 6.11 The Browser view showing the Sidebar on the left and Content Pane on the right

Typically the Browser is used in one of two ways: manually browsing or searching for files.

To manually browse for files after opening the Browser:

- Use the shortcuts in the Sidebar on the left to browse any of the following:
  - Your tagged Collections
  - Ableton's predefined Categories such as Instruments and Plug-Ins
  - Places such as installed Packs, your User Library, the Current Project, and any other folders that you've manually added to the Browser

Figure 6.12 The Sidebar in the Browser

To search for files using standard Browser functionality:

1. Press **COMMAND+F** (Mac) or **CTRL+F** (Windows).

2. (Optional) Select a shortcut from the Sidebar to restrict the search.

   If no shortcut is selected, the search will default to "All results" and return results from all Collections, Categories, and Places.

3. Type a search term into the Browser Search Field.

File Drag and Drop Locations

Once you've located a file on your desktop or using the Browser, you'll want to consider your options for dragging and dropping the file into Ableton Live. Imported files can be dragged to a new track using the Clip/Device Drop Area, or onto an existing track.

To import a file to a new track:

1. Select an audio or MIDI file in the Browser, the Mac Finder, or Windows Explorer.

2. Drag the file into the Clip/Device Drop Area in the Arrangement View or Session View. (See Figure 6.13.) A new track will be added to the project containing the imported audio or MIDI data.

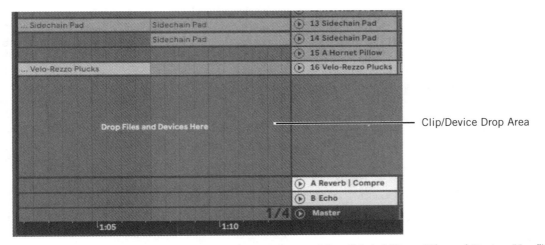

Figure 6.13 The Clip/Device Drop Area in the Arrangement View (labeled "Drop Files and Devices Here")

To import a file to an existing track:

1. Select an audio or MIDI file in the Browser, the Mac Finder, or Windows Explorer.

2. Drag and drop the file to the desired location on an existing track.

Using the Edit Grid

Before you start editing clips, you should have a good understanding of how to work with Ableton Live's Edit Grid. The Edit Grid settings impact the movement and placement of clips and the function of several edit commands. You should also know how to modify the grid value so that clips will align properly when working with Snap to Grid enabled.

Edit Grid Modes

Ableton Live's editing grid can be set to one of two modes: fixed grid and zoom-adaptive grid.

Fixed Grid

In fixed grid mode, selections, clip movements, and resizing operations are constrained to a specific grid value that stays consistent at any zoom level.

Zoom-Adaptive Grid

In zoom-adaptive grid mode, selections, clip movements, and resizing operations are constrained to a grid value that changes based on the zoom level. Zooming in causes the Edit Grid to switch to progressively smaller values.

Setting the Edit Grid Mode and Value

To switch between edit grid modes and values, do one of the following:

- With no clip selected, **RIGHT-CLICK** in the Arrangement View and select the desired edit grid mode and value from the context menu:

    - Select an **ADAPTIVE GRID** option, from **WIDEST** to **NARROWEST**, to enable zoom-adaptive grid mode.

        The **WIDEST** setting uses grid values starting at 16 bars when fully zoomed out; the **NARROWEST** setting uses grid values starting at 1/2 note when fully zoomed out.

    - Select a **FIXED GRID** option, from **8 BARS** to **1/32 NOTE**, to enable fixed grid mode.

    - Select **OFF** under **FIXED GRID** to disable the Edit Grid.

    > ⓘ **Disabling the Edit Grid allows you to position clips freely anywhere within the Arrangement View, without being constrained to grid lines.**

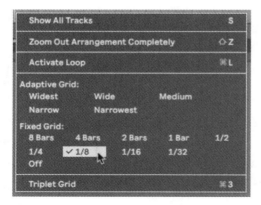

Figure 6.14 The Arrangement View context menu showing edit grid options

- Press **COMMAND+5** (Mac) or **CTRL+5** (Windows) to toggle between fixed and zoom-adaptive grid.

Clip Editing in Arrangement View

Once you've recorded or imported some audio or MIDI material, it's time to start editing!

Basic Editing Techniques

DAWs provide a variety of ways to slice and dice audio and MIDI clips. You'll become familiar with a multitude of editing techniques in Ableton Live as you work, but for now we'll cover the basics techniques you'll need to get started.

Here is basic list of editing techniques:

- Selecting clips
- Moving clips
- Cutting, Copying, and Pasting clips
- Deleting clips

These are the basic building blocks that you'll want to learn. Let's go through each in some detail.

Making Selections

Selecting material is one of the most basic and frequently used techniques in the DAW world. It serves as a building block for further editing. Many other commands require a clip (or portion of a clip) to be selected first before they can be executed. Ableton Live allows you to select an entire clip or just a portion of a clip without changing tools.

To select an entire clip:

- Click once in the upper half of the clip with the hand icon.

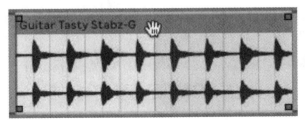

Figure 6.15 Selecting an entire clip by clicking in the upper half

To select time within a clip:

- Click and drag (forwards or backwards) in the lower half of the clip with the pointer icon.

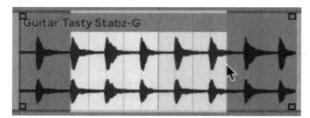

Figure 6.16 Selecting time within a clip by clicking and dragging in the lower half

 Selections can include multiple clips, partial clips, and blank areas between clips. Edit operations will utilize the entire selection and preserve blank areas, as appropriate.

Moving Clips

At times you may want to move a clip to a new location on a track or onto a completely different track. While you can use Cut/Copy/Paste for this purpose, it's usually faster to simply drag the clip to the desired location. Just like selecting, Ableton Live allows you to move an entire clip or just a portion of a clip without changing tools.

To move an entire clip:

- Click in the upper half of the clip and, without releasing the mouse button, drag the clip to the left or right.

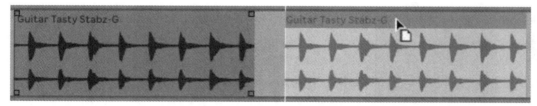

Figure 6.17 Moving an entire clip by clicking and dragging in the upper half

To move a time selection within a clip:

1. Make the time selection by clicking and dragging (forwards or backwards) in the lower half of the clip with the pointer icon.

2. Release the mouse button.

3. Click in the upper half of the clip within the existing time selection and, without releasing the mouse button, drag the selected portion of the clip to the left or right.

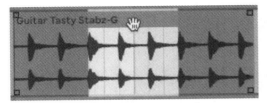

Figure 6.18 Clicking in the upper half of an existing time selection within the clip

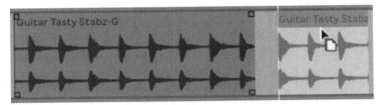

Figure 6.19 Moving the time selection

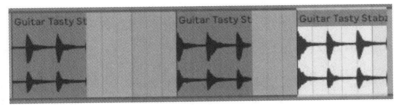

Figure 6.20 The result of moving the time selection

(i) Dragging a clip vertically will move the clip to another track in your set.

(i) Clip movement in the Arrangement View may be constrained, depending on the current grid setting.

Cutting, Copying, and Pasting Clips

For edits involving partial clips or long selections, you'll often want to use Cut/Copy/Paste commands. These commands can be accessed at the top of the Edit menu in Ableton Live, but you'll definitely want to use the keyboard shortcuts to save time.

Ableton Live uses standard keyboard shortcuts for each of these operations:

- Cut operation—**COMMAND+X** (Mac) or **CTRL+X** (Windows)
- Copy operation—**COMMAND+C** (Mac) or **CTRL+C** (Windows)
- Paste operation—**COMMAND+V** (Mac) or **CTRL+V** (Windows)

> ### Duplicate Command
>
> The Duplicate command provides a way to quickly create multiple instances of a clip or selection without copying and pasting.
>
> To use the Duplicate command, select a clip or a portion of a track. Then choose **EDIT > DUPLICATE** or press **COMMAND+D** on the Mac (**CTRL+D** on Windows). A copy of the selected material will be placed immediately after the selection. The Duplicate command can be applied multiple times to create a repeating phrase or pattern.

Deleting Clips

You can remove clips or portions of clips using the **DELETE** or **BACKSPACE** key on the keyboard.

To delete a clip or time selection within a clip:

1. Click in the upper half of a clip to select it, or make a time selection within a clip by clicking and dragging in the lower half of the clip.
2. Press the **DELETE** key (Mac) or **BACKSPACE** key (Windows). The selected clip (or portion of a clip) will be removed from the track.

Advanced Editing Techniques

Beyond selecting and moving your audio or MIDI data, you'll often need to use slightly more advanced editing techniques. Some useful options include resizing the start/end of a clip, splitting a clip into two or more separate clips, and adding fades to clips. All of these operations use non-destructive editing in Ableton Live.

Non-Destructive Audio Editing

Non-destructive editing is an important concept to understand when editing audio data in a DAW. Essentially, non-destructive editing means that changes you make to an audio clip will not affect the original audio file stored on disk. This gives you the freedom to try any number of different edits without fear of permanently modifying the original file.

Editing MIDI Clips

While editing audio data is almost always non-destructive, the same is not true for MIDI data. This is because MIDI clips in Ableton Live are stored within the project file, and not in a separate file on disk. As a result, some MIDI editing commands *will* modify the original clip. Although it is possible to use the undo command to revert a clip back to the original, this option is typically useful only immediately after making an unwanted change.

Resizing a Clip

Ableton Live offers a clip-resizing feature that can be used to adjust clip boundaries. This operation lets you hide or expose underlying material. The resizing feature can be used on audio, MIDI, and video clips. Like all clip editing in Ableton Live, resizing a clip is *nondestructive*, leaving the underlying media file unchanged.

To resize a clip:

- Click and drag in the upper half of the clip near the clip boundary (beginning or end) where the bracket cursor appears.

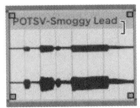 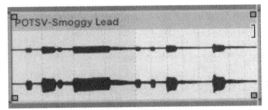

Figure 6.21 Extending the end of a clip to expose underlying material from the file; before (left) and after (right)

Note that when extending the length of a clip, there are two possible outcomes:

- If the clip has underlying material available in the source media file, the clip will extend normally.
- If there is not additional material available, Ableton Live will automatically loop the clip to create the desired extension.

Splitting a Clip into Two or More Clips

Another very common editing technique is to split a clip into two or more smaller clips. In Ableton Live, you can use a simple edit command to perform this function: **EDIT > SPLIT**.

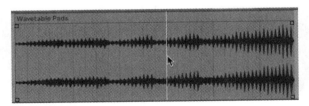 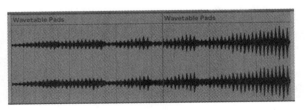

Figure 6.22 A single clip before splitting (left) and the result of splitting at the insert marker (right)

To split a clip into two clips:

1. Click at the desired location in the bottom half of a clip to position the Insert Marker.

2. Do one of the following:

   - Select **EDIT > SPLIT**.

   - Press **COMMAND+E** (Mac) or **CTRL+E** (Windows).

Fading Clips

Last but certainly not least, you can apply fades to your audio clips. Fades are not only used to gradually fade audio in and out, but also to prevent undesirable pops and clicks at the beginning and end of clips. Here, we'll focus on the basics of creating and editing fades in Ableton Live.

To create a fade in or fade out, do one of the following:

- Click and drag the fade handle in the upper corner of the clip.

- Select the beginning or ending of a clip and choose **CREATE > CREATE FADE IN/OUT** or press **OPTION+COMMAND+F** (Mac) or **CTRL+ALT+F** (Windows).

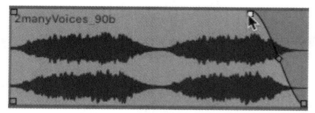

Figure 6.23 Using the fade handle to create a fade out on an audio clip

In addition to fade–ins and fade–outs, the Fades operation can be used to create crossfades. Crossfades are useful to smooth out an edit location, by fading out the audio from before the edit point while simultaneously fading in the audio from after the edit point.

To create a crossfade using fade handles:

- Click the fade in or fade out handle and drag it across the opposite clip's edge.

- Select a range of time that includes the boundary between adjacent clips and choose **CREATE > CREATE CROSSFADE** or press **OPTION+COMMAND+F** (Mac) or **CTRL+ALT+F** (Windows). (See Figure 6.24)

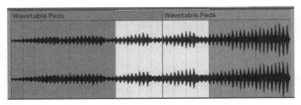 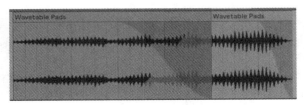

Figure 6.24 Two adjacent clips: before creating a crossfade (left) and after creating a crossfade (right)

If you decide to modify the fades later, you can use the fade handles to resize the fade or to modify the fade curve.

MIDI Editing Techniques

MIDI editing is another topic worthy of its own chapter (or several chapters). But as with audio editing, a little knowledge is all you need to get started. Understanding the fundamentals of MIDI editing is absolutely essential in the modern studio. Whether your goal is to add a few MIDI instruments to a song or to compose a virtual orchestral masterpiece, you'll need to know how to edit MIDI data quickly and accurately.

Editing MIDI Clips in the Arrangement View

MIDI clips can be edited in the Arrangement View using the same editing techniques you use for audio clips:

- Cutting, copying, pasting, and deleting selections or clips
- Duplicate selections or clips
- Trimming clips

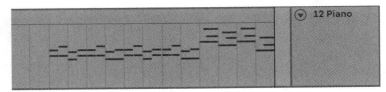

Figure 6.25 A MIDI clip displayed in the Arrangement View

Using the MIDI Note Editor

Live's MIDI Note Editor allows you to both create new MIDI performances and edit existing ones. To access Live's MIDI Note Editor, select a MIDI clip (if Clip View is active) or double-click a MIDI clip (if Clip View is not active). This will display the clip contents in Clip View, within the MIDI Note Editor.

Click and drag the divider between the Clip View and Track Display to expand the MIDI Note Editor, if necessary.

The MIDI Note Editor shows individual MIDI notes in a piano-roll format, with note pitch shown on the vertical axis, and note position and duration in time shown on the horizontal axis. MIDI notes are represented as colored bars of different brightness, depending on their respective velocities. (See Figure 6.26.)

A dedicated velocity editor is located along the bottom of the MIDI Note Editor, with the current grid spacing value indicated in the lower right-hand corner.

 If the velocity editor is not visible, click the triangle icon at the bottom of the note ruler area in the MIDI Note Editor (bottom left corner).

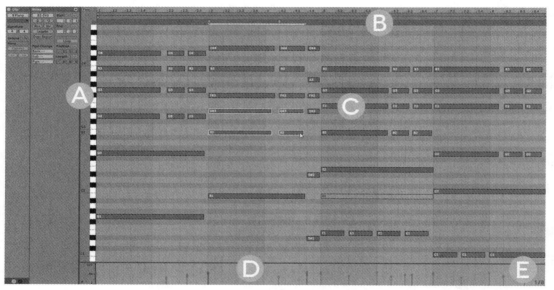

Figure 6.26 The MIDI Note Editor: (A) note ruler (B) time ruler (C) MIDI notes (D) velocity editor (E) grid spacing

The visual nature of the MIDI Note Editor makes it easy to see musical performances as well as to create, select, and edit notes. The sections below cover editing functions that will allow you to create the exact performance you desire.

Creating MIDI Notes

During the production process, you may want to enter MIDI notes by hand. For many producers, this is a faster way of entering MIDI data than playing a keyboard or other controller.

Live provides two methods of creating MIDI notes by hand:

- Double-click in the MIDI Note Editor at the desired note and position in time. Before releasing the mouse, drag to the right to extend the duration of the note as needed. (You can also trim note durations after adding them by dragging on a note edge with the trim icon.)

- Press the **B** key or the **DRAW MODE** button in the Control Bar to enable Draw Mode. This enables the pencil tool, allowing you to "draw" MIDI notes in the editor. Note lengths will match the current grid size of the MIDI Note Editor.

Figure 6.27 The Draw Mode button (yellow) in the Control Bar

When adding notes using the above methods, note start and end points will snap to the current grid lines. To change the grid size, right-click in the MIDI Note Editor and choose a different Fixed Grid setting. You can also press and hold **COMMAND** (Mac) or **ALT** (Windows) to remove grid constraints when adding new notes.

 With the cursor in the MIDI Note Editor, press **COMMAND+1** and **COMMAND+2** (Mac) or **CTRL+1** and **CTRL+2** (Windows) to narrow or widen the grid, respectively.

 Clicking on an existing note in Draw Mode (or double-clicking in Edit Mode) will delete the note.

Selecting MIDI Notes

Whether you have recorded or manually created MIDI notes, you can make selections for editing in multiple ways.

To make a selection of MIDI notes, do one of the following:

- Click an individual note in Edit Mode.

- Starting from an empty space, click and drag within the MIDI Note Editor in Edit Mode to draw a marquee around multiple notes.

- **SHIFT+CLICK** a note in Edit Mode to add a note to a selection or remove a note from a selection.

- Click and drag within the MIDI Note Editor in Edit Mode *without* selecting any notes to select a timespan, or click and drag within the MIDI Note Editor Scratch Area in either mode to create a timespan. Then press **RETURN** or **ENTER** to select all notes that start within the timespan.

- With the cursor located in the MIDI Note Editor or with any note selected, press **COMMAND+A** (Mac) or **CTRL+A** (Windows) to select all notes in the clip.

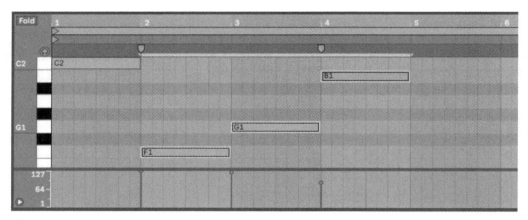

Figure 6.28 A multi-note selection in the MIDI Note Editor

Editing MIDI Notes

The MIDI Note Editor provides many ways to edit notes, allowing you to develop your own workflows. It is possible to edit MIDI notes using a mouse or trackpad.

You can quickly change the pitch, position, and length of MIDI notes using the mouse.

To edit MIDI notes using a mouse or trackpad:

- Click and drag a note in Edit Mode to move it to another pitch and/or position in time.
- Click and drag the edges at the start or end of a note to trim its duration in either direction.
- Right-click and select **Deactivate Note(s)** to prevent a note from playing (without deleting it).
- Right-click and select **Duplicate** to duplicate a note or a selection of notes.
- While holding **Option** (Mac) or **Ctrl** (Windows), click and drag a note or selection of notes to duplicate it.

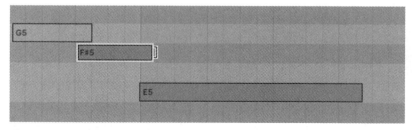

Figure 6.29 A deactivated MIDI note (G5) and a note being trimmed with the mouse (F#5)

Editing Velocity

Velocity is MIDI data that indicates how hard a key or pad was pressed when recording from a MIDI controller. This data is stored within the MIDI performance. Velocity information can help you create more lifelike MIDI performances, as different velocity values can trigger different audio samples and different volume levels from virtual instruments. Even if you don't perform MIDI using a velocity-sensitive MIDI controller, you can manually edit velocity data for each note in a clip.

Velocity is represented both by the brightness of a note and by the velocity stalk at the bottom of the MIDI Note Editor. Higher velocity notes are brighter and more vibrant in color. Their velocity stalk in the velocity editor is also taller than lower-velocity notes.

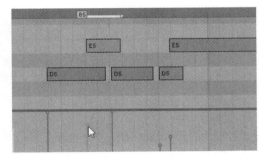

Figure 6.30 Editing velocity by clicking and dragging a velocity stalk

To quickly adjust the velocity for one or more notes:

1. Select one or more MIDI notes.
2. Do one of the following:
   - Click and drag a velocity stalk up and down in the velocity editor.
   - While holding **COMMAND** (Mac) or **ALT** (Windows), click and drag up and down on a MIDI note.

The velocity value will display at the top of the MIDI Note Editor window, using a range spanning from 1 to 127. When multiple notes are selected, the velocity of all notes will change relative to one another.

Navigating the MIDI Note Editor

Depending on the complexity of your MIDI performance, it may be very useful to zoom the MIDI Note Editor to increase the visual resolution of the recorded notes. You can navigate and zoom the vertical and horizontal axes independently, allowing you to isolate specific notes and make fine-tuning adjustments, or show all of the MIDI data in a clip for broad edits.

Scrolling and Zooming with Time Ruler (Horizontal Axis)

The horizontal axis, or time ruler, represents the position and duration of MIDI notes. There are multiple ways to zoom the time ruler for specific editing tasks.

Figure 6.31 The cursor will display a magnifying glass when positioned over the time ruler

To zoom horizontally using the timer ruler:

- Click and drag up and down on the time ruler to zoom the MIDI Note Editor horizontally.
- Double-click the time ruler to show the contents of the entire selected MIDI clip.
- With one or more notes selected, double-click the time ruler to fit the selection horizontally to the MIDI Note Editor window.

> ⓘ With the cursor placed in the MIDI Note Editor (or with a note selected), you can also press the PLUS (+) and MINUS (–) keys on the keyboard to zoom horizontally.

Once you have zoomed in, you can click and drag left and right on the time ruler to scroll the MIDI Note Editor horizontally. You can also scroll left and right from the keyboard by pressing **COMMAND+PAGE UP/PAGE DOWN** (Mac) or **CTRL+PAGE UP/PAGE DOWN** (Windows).

Scrolling and Zooming with the Note Ruler (Vertical Axis)

The vertical axis, or note ruler, represents all the notes or pitches that you can send to a MIDI instrument. This display format is known as a piano roll. Ableton Live provides multiple ways to zoom using the note ruler for specific editing tasks.

Figure 6.32 The cursor will display a magnifying glass when positioned over the note ruler

To zoom vertically using the note ruler:

- Click and drag left and right on the note ruler to zoom the MIDI Note Editor vertically.
- Double-click the note ruler to show the entire range of notes used in the selected clip.

- With multiple notes selected, double-click the note ruler to fit the selection vertically to the MIDI Note Editor window.

> ⓘ When zooming in vertically on selected notes, the range will scroll to begin at the top of the MIDI Note Editor. If the range of notes does not fit at maximum vertical zoom, additional pitches will be displayed below the selected range.

Once you have zoomed in, you can click and drag up and down on the note ruler to scroll vertically through the octaves of available notes. You can also scroll up and down from the keyboard using the **Page Up** and **Page Down** keys.

> ⓘ Click the **Fold** button in the upper left-hand corner of the MIDI Note Editor to hide any MIDI pitches that are unused in the selected clip.

Review/Discussion Questions

1. How can you specify a starting point (punch in location) for recording? (See "Designating a Punch-In Point" beginning on page 139.)

2. What are two ways that you can begin a record take in Ableton Live? (See "Initiating a Record Take" beginning on page 141.)

3. What are two ways that you can stop a record take? (See "Stopping a Record Take" beginning on page 141.)

4. How can you immediately delete a recorded clip? (See "Stopping a Record Take" beginning on page 141.)

5. How is MIDI Arrangement Overdub useful when recording drum parts? (See "Using MIDI Arrangement Overdub" beginning on page 143.)

6. What are some types of audio files that can be imported into Ableton Live? (See "Supported Audio Files" beginning on page 144.)

7. What two locations are available as destinations for files you import using the drag–and–drop method? (See "File Drag and Drop Locations" beginning on page 147.)

8. What are the two Edit Grid modes in Ableton Live? What is the difference between them? (See "Edit Grid Modes" beginning on page 147.)

9. How can you switch between the Edit Grid modes? How can you change the Edit Grid value? (See "Setting the Edit Grid Mode and Value" beginning on page 148.)

10. How can you select a portion of a clip in the Arrangement View? (See "Making Selections" beginning on page 149.)

11. Where must you click in a clip to drag and drop it to a new location in the Arrangement View? (See "Moving Clips" beginning on page 150.)

12. Is audio editing generally a destructive process or non-destructive process in Ableton Live? What about MIDI editing? (See "Advanced Editing Techniques" beginning on page 152.)

13. What Ableton Live command is used to split a clip into two separate clips? (See "Splitting a Clip into Two or More Clips" beginning on page 153.)

14. What are some edit techniques that can be performed on MIDI clips in the Arrangement View? (See "Editing MIDI Clips in the Arrangement View" beginning on page 155.)

15. How does MIDI note data appear in the MIDI Editor? (See "Using the MIDI Note Editor" beginning on page 155.)

16. What two methods can be used to create MIDI notes by hand in the MIDI Editor? (See "Creating MIDI Notes" beginning on page 156.)

To review additional material from this chapter and prepare for certification, see the Ableton Live Audio Production Basics Study Guide module available through the Elements|ED online learning platform at ElementsED.com.

Exercise 6

Importing and Editing Clips

🎧 Activity

In this exercise, you will perform various importing and editing techniques in Ableton Live. You'll begin by opening an existing project. Then you'll import an audio file and a MIDI file into the project and configure the tracks. Finally, you'll do some basic audio editing to remove unnecessary content from two audio tracks.

⏱ Duration

This exercise should take approximately 15 minutes to complete.

◎ Goals/Targets

- Open an existing project document
- Import an audio file into the project
- Import MIDI into the project
- Assign a virtual instrument for the Piano track
- Remove audio at the beginning of two drum tracks for a 4-bar intro

---

Exercise Media

This exercise uses media files taken from the song, "Lights," provided courtesy of Bay Area band Fotograf.

Written by: Zack Vieira and Eric Kuehnl; Performed by: Fotograf

The media provided for this course may be used for educational purposes only. No rights are granted to use the media for any other personal, commercial, or non-commercial purposes.

Importing and Editing Clips **165**

Getting Started

To get started, you will launch Ableton Live and then open the exercise Ableton Live Set. If you are running Ableton Live Lite, you can use an alternate version of this Set designed for a lower track count.

Launch Ableton Live:

1. Launch Ableton Live. A new Ableton Live Set will be created and will display on screen.
2. Choose **FILE > OPEN LIVE SET**. The Open Document dialog box will appear.
3. Navigate to your **Documents** folder or other location where you saved the **Live APB Media Files** folder in Exercise 1; open the **06-Lights** folder within the **Live APB Media Files** folder.
4. Double-click on the **Lights.als** Set file to open it. (If you are running Live Lite, open the **LightsLite.als** Set file instead.) The Live Set will open.

> If one or more samples are missing (offline), the Missing Files warning will appear in red at the bottom of the Ableton Live window. Click the warning message to resolve the issue.

> See Appendix A for details on locating missing files.

5. Choose **FILE > SAVE LIVE SET AS**. The Save Live Set As dialog box will appear. If needed, press **TAB** to switch to Arrangement View.
6. Name your Set **Lights01-*xxx***, where *xxx* is your initials, and navigate to an appropriate save location.

> In a classroom or lab environment, you may be required to save or submit your work using a specific location. Check with your instructor for details.

7. Click the **SAVE** button at the bottom of the Save Live Set As dialog box to save the Set.

Import Audio

The exercise Set already includes audio clips for most of the tracks. However, the lead vocal clip has not been added yet, so you will need to import that into the project. For this process, you will use the Browser sidebar and point it to the folder containing the file you need to import.

> Prior to completing this section, verify that the Preference setting for Auto-Warp Long Samples is set to OFF. (Choose **LIVE > PREFERENCES** on Mac or **OPTIONS > PREFERENCES** on Windows and select the **RECORD/WARP/LAUNCH** tab.)

Add a folder to the Browser sidebar:

1. In the Browser sidebar, locate the Places section and click **ADD FOLDER**.

2. In the resulting dialog box, navigate to your Documents folder (or other location where you saved the Live APB Media Files folder in Exercise 1).

3. Open the Live APB Media Files folder.

4. Select the Media To Import folder and click **OPEN**. The folder will now appear in the Places area of the Browser sidebar. This will let you easily access the files you need to complete the project.

Import the lead vocal audio:

1. Select the Media To Import location you added under Places in the Browser sidebar. Then select the Vox.wav file under the Name column in the Browser sidebar.

2. Drag and drop the file to the start of a new track, using the Clips/Device Drop Area in the Arrangement View. A track named Vox will be added to the project containing the imported audio.

> You can recognize the Clips/Device Drop Area by the displayed text label, "Drop Files and Devices Here" above the Master track.

3. Click the Optimize Arrangement Height (**H**) and Optimize Arrangement Width (**W**) buttons in the upper right corner of the Arrangement View to fit the tracks to the screen height and width.

When finished, your project should look similar to Figure 6.33 below.

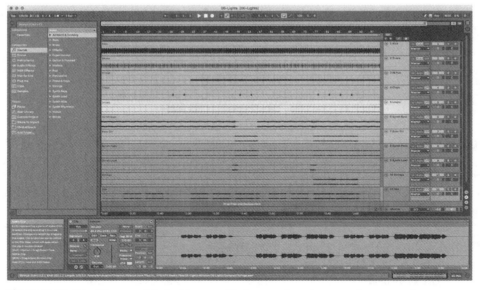

Figure 6.33 The Live set with the Vox.wav file imported to the start of a new track

Import MIDI

In this part of the exercise, you will import a MIDI clip contained in a Standard MIDI file.

Import the MIDI data:

1. Use the Browser sidebar to select the **Ex06.mid** file in the **Media To Import** folder.
2. Drag and drop the file into the Clip/Device Drop Area at the bottom of the Arrangement View.
3. A dialog box will appear asking if you want to import tempo and time signature data into the Arrangement. Click **YES**.

Figure 6.34 The Import Tempo and Time Signature dialog box

A new MIDI track will be added to the project containing the imported MIDI data.

Rename the MIDI track:

1. Select the new MIDI track and press **COMMAND+R** (Mac) or **CTRL+R** (Windows) to bring up the Rename dialog box.
2. Rename the track as **Piano**.

Assign a Virtual Instrument

With the MIDI clip imported to a MIDI track, you'll next need to assign a virtual instrument to that track. You'll use the Simpler virtual instrument device (included with all editions of Ableton Live) on the track.

Insert Simpler on the MIDI track:

1. Under the **Categories** section of the Browser sidebar, navigate to the following location: **Instruments > Simpler > Piano & Keys > Grand Piano.adg**. (See Figure 6.35.)

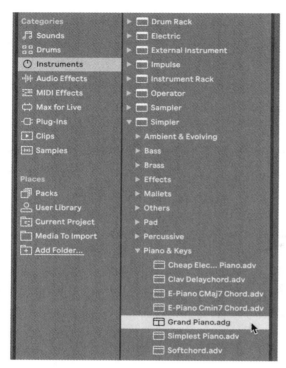

Figure 6.35 Selecting the Grand Piano.adg preset for Simpler

2. Drag and drop the **Grand Piano.adg** preset file onto the **Piano** track. The virtual instrument will appear in the Device view at the bottom of the main window.

Figure 6.36 The Grand Piano preset

 The Grand Piano.adg preset is actually an Instrument Rack. As a result, the controls displayed in the Device view are the Instrument Rack's Macro controls and not the standard Simpler interface.

Preview the project:

1. Press the **SPACEBAR** to begin playback and listen to the project in its current state.

2. Pay particular attention to the number and type of instruments that you hear. Feel free to solo tracks (with the **S** button) and/or deactivate tracks (with the **TRACK ACTIVATOR** switch) to isolate certain elements and get familiar with the different parts.

Remove Unwanted Audio

In this part of the exercise, you will remove some drum parts at the beginning of the song. This will create a more dramatic intro featuring the Synth Bass part.

Remove the unwanted audio:

1. Click near the start of the **Kick** track and press the plus (**+**) key several times to zoom in on the first few measures of the song.

2. Choose **OPTIONS > SNAP TO GRID** or press **COMMAND+4** (Mac) or **CTRL+4** (Windows), if needed, to enable **SNAP TO GRID**.

> ⓘ The Snap To Grid function toggles on/off with the menu command or the keyboard shortcut. When enabled, the vertical Grid lines will appear solid in the Arrangement View.

3. Set the Edit Grid value by right-clicking in the Arrangement View and selecting **FIXED GRID: 1 BAR** from the pop-up menu. This will ensure that all of your selections are in precise increments of one bar.

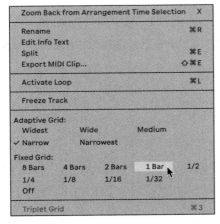

Figure 6.37 Setting the Edit Grid value to Fixed Grid: 1 Bar

4. Click and drag in the bottom half of the **Kick** clip to select the first four bars (1.1.1 to 5.1.1).

5. Press **DELETE** to delete the selection.

6. Repeat Steps 4 and 5 for the **Snare** track.

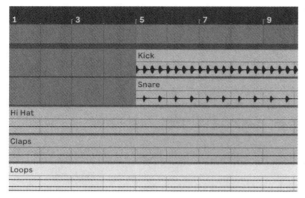

Figure 6.38 The first four bars removed from the Kick and Snare tracks

Finishing Up

To complete this exercise, you will need to save your work and close the project. You will be reusing this project in Exercise 8, so it's important to save and retain the work you've done.

Before you wrap up, you should also listen to the project to hear the changes you've made.

Finish your work:

1. Press **STOP** in the transport section of the Control Bar to position the Insert Marker at the beginning of the project.

2. Press the **SPACEBAR** to begin playback. Only the synth bass will be audible for the first four bars, creating a nice intro.

3. When finished auditioning, press the **SPACEBAR** a second time to stop playback.

4. Choose **FILE > SAVE LIVE SET** to save the Set.

5. If desired, exit Ableton Live by doing one of the following:

   - On a Mac-based system, choose **LIVE > QUIT LIVE**.
   - On a Windows-based system, choose **FILE > EXIT**.

   Ableton Live will automatically close.

That completes this exercise.

Chapter 7

Session View

...How to Make Use of the Ableton Live Session View...

This chapter introduces you to some basic operations and functions you need to work with Ableton Live's Session View. We begin with an overview of the sections and controls within the Session View. Then we take a look at techniques for adding clips to tracks in the Session View. Next, we look at ways to control playback in the Session View. We conclude with an overview of using the Session View to record audio and MIDI.

⊕ Learning Targets for This Chapter

- Become familiar with the Sections and Controls in the Session View
- Add Clips to Tracks in Session View
- Control playback in Session View
- Record audio in Session View
- Record a MIDI performance in Session View
- Record from Session View into an arrangement

 Key topics from this chapter are illustrated in the Ableton Live Audio Production Basics Study Guide module available through the Elements|ED online learning platform. Sign up for a free account at ElementsED.com.

The Session View

This chapter introduces you to the basic operations and functions you need to work with Ableton Live's Session View. The Session View is one of the most unique aspects of Ableton Live. Although its behavior is unusual among DAWs, the Session View offers an efficient set of tools for both composition and live performance. Once you understand the basic operation, the functionality becomes quite intuitive and easy to use.

When you're working in the Session View, you will have access to a series of scenes and clips. You can launch clips and scenes in any order to generate performances without being tied to a linear timeline.

Figure 7.1 The Session View is displayed beneath the Control Bar (also shown: Browser View and Detail View).

The Session View provides a number of controls grouped into sections. Several of these sections, such as the Clips Slots section and the Scenes section, are unique to the Session View. Other sections, such as the In/Out, Sends, and Mixer sections, can also be found in the Arrangement View.

 You can toggle between the Session View and Arrangement View by pressing the Tab key.

Session View Clip Sections

The Clip Slots and Scenes sections of the Session View are dedicated to organizing and launching clips for composition and live performance.

ⓘ Ableton Live Intro permits up to eight clip slots per track and scenes across a maximum of 8 tracks. Ableton Live Standard and Suite support an unlimited number of clips slots per track and scenes.

ⓘ In Session view, tracks are represented by vertical columns, which are similar to traditional mixer strips (aside from the Clip Slots at the top). Scenes are represented by horizontal rows in the Clips Slots section.

Clip Slots Section

The Clips Slots section of the Session View offers a collection of slots into which audio and MIDI clips can be placed. When a clip is played (or "launched") from a Clip Slot, the signal from the audio clip or MIDI instrument plays through the signal path comprised of the track sections that follow it (see below).

Figure 7.2 Clip Slots section of the Session View (seven tracks shown)

To launch a clip in a Clip Slot, click on the triangular play button at its left edge. To stop clip playback, press the spacebar.

📖 Detailed information on controlling playback from Clip Slots is provided in the section on "Starting and Stopping Playback" later in this chapter.

Scenes Section

The Scenes section of the Session View is displayed on the Master track and allows you to launch an entire row of clips (a "scene") all at once. (See Figure 7.3.) Scenes are typically used to navigate through sections of a single song (verse, chorus, etc.) during production, or to move between complete songs during performance.

Figure 7.3 Scenes section of the Session View with the Intro scene selected

To launch a scene, click on the triangular play button to the left of the scene name. To stop scene playback, press the spacebar.

 Detailed information on controlling scene playback is provided in the section on "Controlling Playback with Scenes" later in this chapter.

To rename a scene:

- Right-click on a scene name on the Master track and choose **RENAME** from the pop-up menu. The scene name will become highlighted, allowing you to type in a new name.

To add a scene to the current Set:

- Choose **CREATE > INSERT SCENE** or press **COMMAND+I** (Mac) or **CTRL+I** (Windows). The scene will be inserted after the currently selected Scene.

To delete a Scene:

1. Select the target Scene by clicking on the scene name on the Master track.
2. Choose **EDIT > DELETE** or press the **DELETE** key on the computer keyboard.

To reposition a Scene:

- Click on a scene name on the Master track and drag it to a new position in the Scenes section.

Session View Mixing Sections

The Session View also includes sections and controls that are used for signal routing and mixing. Among the controls that you will use to create your mix are the following (see Figure 7.4):

- I/O choosers in the In/Out section
- Send controls in the Send section
- Pan controls in the Mixer section
- Volume controls in the Mixer section

Although all of these mixing controls can be viewed in the Arrangement View, mixing operations and functions are typically performed using the Session View. The Session View is similar to a standard mixing console. This view offers a variety of display options, many of which can also be customized.

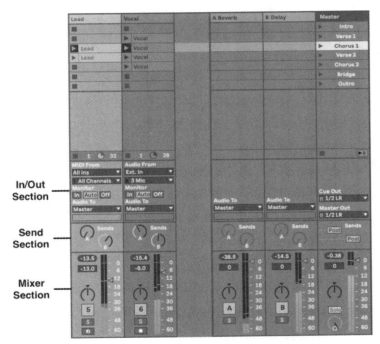

Figure 7.4 Mixing controls in the Session View

In/Out Section

The top portion of each channel strip in the Session View provides controls for routing signals for the track. Depending on the type of track, these controls may include Audio Input and Output selectors, MIDI Input and Output selectors, and Monitor controls. (See Figure 7.5.)

Figure 7.5 The In/Out section of the Session View

Audio Input and Output selectors are used to route input and output signals from your audio interface for recording or playback. MIDI Input and Output selectors are used to route MIDI data from MIDI controllers to either instrument plug-ins or external MIDI hardware. The Monitor controls determine whether you will hear the live track input or playback of an existing clip on a given track.

As discussed in Chapter 6, Monitor controls can generally be left on the **AUTO** setting, which will automatically switch from monitoring live input when a track is recording (or record-armed in stop) to monitoring clip playback when the track is in play.

Send Section

The Session View has a dedicated section for the track send controls known as the Send section. This area can be shown or hidden in the Session View (see "Showing/Hiding Sections" below).

Figure 7.6 Send section in the Session View

The Send section allows you to access and view up to 12 Send Controls (A through L) in Ableton Live Standard and Suite editions. The default Ableton Live Set is pre-configured with two sends (A and B). Additional send controls are added as you create additional return tracks in your Set.

 The Send section is part of the Mixer Section in the Arrangement View.

Mixer Section

The Mixer section of the Session View provides a mixer-like environment for working with tracks. In the Mixer section, tracks appear as mixer strips (also called *channel strips*). Each track displayed in the Mixer section has controls for panning and volume. The Mixer section also provides buttons for enabling record, activating/deactivating tracks, and toggling solo on and off. (See Figure 7.7.)

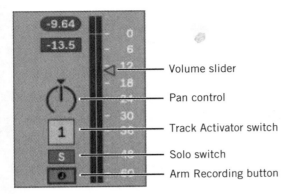

Figure 7.7 The Mixer section of the Session View

The Pan controls can be used to position the output of a track within the stereo field. The Track Activator, Solo, and Arm Recording buttons can be used for mute, solo, and record arm functions, respectively, during record and playback. The Volume slider can be used to adjust the playback/monitoring level of a track.

 The Volume slider in the Mixer Section does not affect the input gain (record level) of a signal being recorded. The signal level must be set appropriately at the source or adjusted using a preamp or gain-equipped audio interface.

Showing/Hiding Sections

You can show or hide sections of the Session View using the Show/Hide buttons at the bottom right corner of the Session View. Click a button to toggle the show/hide status of the associated section.

Figure 7.8 The Session View Show/Hide buttons for the In/Out section

The Show/Hide buttons shown in Figure 7.8 are as follows, from top to bottom:

- I•O – In/Out Section
- S – Send Section
- R – Return tracks

- M – Mixer Section
- D – Track Delays Section
- X – Crossfader Section

Adding Clips to Tracks in Session View

There are several ways that clips can be added to tracks in the Session View. These include importing to existing tracks, importing to new tracks, and batch importing.

Import to Tracks in Session View

While the Session View environment is very different from that of the Arrangement View, the process of placing audio files as clips on tracks is very similar.

To place audio from the Browser on an audio track in the Session View:

1. Select the audio file in the Browser.
2. Click and drag the audio file to a clip slot on an audio track. The slot will update to display a play button and the associated clip name.

With a clip added to a clip slot, you can launch the audio file using either the clip launch button (play arrow on the clip slot) or the associated scene launch button (play arrow for the scene on the Master track).

Import Directly to a New Audio Track in Session View

You can skip the step of manually creating a destination track for an audio file, if desired.

To automatically place an audio file on a *new* audio track, drag the file from the Browser to any position *to the right of* the last track (Session View) in your Live Set. Live will create a new audio track at that position containing the clip.

Batch Importing Audio to Tracks in Session View

At times you might find yourself in a situation where you'd like to import multiple audio files simultaneously. In these cases, you can quickly send a set of selected files to one or more tracks. You can import the audio into subsequent clip slots on a single track, or you can place each audio file on a separate track in the same clip slot position.

To place audio files into subsequent clip slots on the same audio track:

1. Select the audio files in the Browser using the methods described in Chapter 6.

2. Click and drag the audio files to a clip slot on an audio track. The clips will be added to successive slots on the track.

Figure 7.9 Simultaneously dragging multiple audio files to the Drums track in the Session View

To place audio files into clip slots on separate tracks:

1. Select the audio files in the Browser using the methods described in Chapter 6.

2. While holding **COMMAND** (Mac) or **CTRL** (Windows), drag the audio files onto the desired clip slot position on the first of a set of successive target audio tracks. The clips will be added to successive audio tracks at the same slot position.

Batch Import Directly to New Audio Tracks

You can also skip the step of manually creating destination tracks when batch-importing audio files. Instead, you can create the tracks automatically upon import.

To place multiple selected audio files on one new audio track:

- Drag the selected files from the Browser to any position to the right of the last track in the Session View in your Live Set.

  A new track will be added at that position, and the audio files will be added to successive clip slots on the track.

To place each selected file on its own new audio track:

- Hold **Command** (Mac) or **Ctrl** (Windows) while dragging multiple files from the Browser to any position to the right of the last track in the Session View in your Live Set. New tracks will be added at that position as needed, with the audio files added at the same clip slot position on each track.

 You can easily recognize the new track drop area by the displayed text label, "Drop Files and Devices Here."

Playback in Session View

While the Arrangement View displays your Live Set along a fixed timeline, the Session View displays a grid of tracks and clips that you can play at any time and in any order. The Session View is designed for improvisation and dynamic, non-sequential playback, so the layout of tracks and clips does not predetermine their playback order.

Starting and Stopping Playback

Each clip that you load in the Session View has a triangular **Clip Launch** button (Play button) at its left edge. Clicking this button will launch playback for the clip at any time. Launching a clip will also activate playback of the Arrangement, if it is not already in progress.

Figure 7.10 The triangular Clip Launch button

Clips will typically play continuously until stopped, in a repeating (looping) pattern.

 For information on controlling loop behavior for clips, see "Enabling/Disabling Clip Looping" later in this chapter.

To stop a clip from playing, click the **Clip Stop** button on any empty Clip Slot in the track or in the Track Status Display. This will cause the clip to stop playback at the end of its cycle; however, the Arrangement will continue to play.

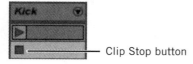

Figure 7.11 Click the square Clip Stop button below an active clip to stop it from playing.

When you stop playback for a specific clip in the Session View, the Play button in the Control Bar will remain engaged and the Arrangement Position counter will continue running. These fields maintain a continuous flow of musical time so that you always know your position in song-time during a live performance or while recording clips into an arrangement.

You can stop the Arrangement playback by pressing the **STOP** button in the Control Bar or by pressing the **SPACEBAR**. Pressing the **STOP** button a second time (or anytime while playback is stopped) will return your Arrangement Insert Marker to the start of the entire Live Set.

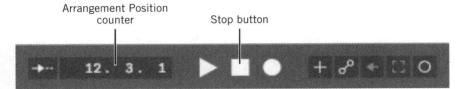

Figure 7.12 The Control Bar, including Play and Stop buttons, as well as the Arrangement Position counter

Controlling Playback with Scenes

The Scene Launch buttons in the Scenes section of the Session View can be used to launch all of the clips in a horizontal row.

To launch a Scene, do one of the following:

- Click on the associated **SCENE LAUNCH** button on the Master track.

- Select the scene by clicking on it on the Master track; then press the **RETURN** key (Mac) or **ENTER** key (Windows).

> If the Select Next Scene on Launch option is enabled in the Launch preferences, the scene below the launched Scene will automatically be selected as the next to be launched. This makes it easy to launch Scenes from top to bottom without having to select them individually.

Setting Tempo and Time Signature Using Scenes

Scenes can also be used to modify the tempo and time signature of your project. If Live detects that part of the scene name includes tempo or time signature information, the project will automatically change the tempo or time signature when the scene is launched. A scene may contain a tempo change, a time signature change, or both.

Figure 7.13 Scenes containing tempo and time signature information (right)

To assign a tempo to a Scene:

- Rename the scene to include the desired tempo value followed by **BPM**. For example, rename **Chorus** to **Chorus 123 BPM**.

To assign a time signature to a Scene:

- Rename the scene to include the desired time signature expressed as a fraction. The numerator must be between 1 and 99, and the denominator must be either 1, 2, 4, 8, or 16. For example, rename **Bridge** to **Bridge 3/4**.

Enabling/Disabling Clip Looping

Live's Session View also provides the ability to enable or disable looping for individual clips. Clips that you import through the Browser will often be Loop-enabled by default. This allows the clip to loop indefinitely when launched in the Session View.

When used the Arrangement View, a Loop-enabled clip can be resized beyond the length of its original media, with Ableton Live automatically adding successive iterations (repetitions) of the clip contents to achieve the desired length.

To enable or disable looping for a clip, do the following:

1. Do one of the following to display the clip in Clip View:
   - If Clip View is not active, double-click on the clip in a Clip Slot (Session View) or double-click in the top part of the clip on a track (Arrangement View).
   - If Clip View is active, click on the clip in a Clip Slot (Session View) or click on the top part of the clip on a track (Arrangement View).
2. Click on the **CLIP LOOP** switch in the Clip View to toggle looping on/off. (See Figure 7.14.)

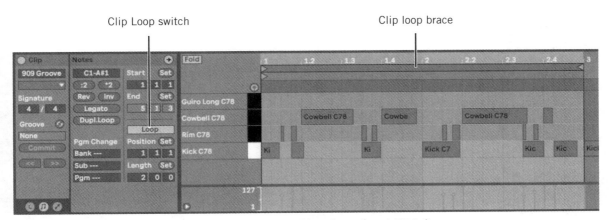

Figure 7.14 The Clip View, showing the Loop switch and loop brace for a MIDI clip

 The Warp switch must be activated to loop an audio clip. Unwarped audio clips cannot be looped.

Recording Audio in Session View

Recording in Session View allows you to create individual clips and launch other clips while recording.

To begin recording audio in Session View, do the following:

1. Arm one or more tracks for recording. Once armed, each clip slot in the track will have its own record button.

2. Click the **RECORD** button on one of the clip slots to begin recording a new clip to that slot.

    The clip slot record button will flash red, and the metronome count-off will begin (if enabled). Once recording has commenced, the record button will turn into a red Clip Launch button. (See Figure 7.15.)

Figure 7.15 Recording audio to a clip slot in the Session View

3. When you have finished your record take, click the **Stop** button in the Control Bar (or press the **spacebar**).

   Alternatively, you can press the red Clip Launch button to continue the playback of the current scene, including the new audio recording.

Recording to Multiple Tracks

If you have multiple tracks armed for recording, you can click the **Session Record** button (hollow circle) in the Control Bar to record onto all armed tracks on a selected scene.

Figure 7.16 The Session Record button (highlighted) in the Control Bar

Session View Recording Quantization

When recording in the Session View, you might want to select a value from the Quantization Menu. (See Figure 7.17.) When the value is set to anything other than **None**, Ableton Live will automatically cut the clip boundaries of your recorded audio so they align with your Live Set. This will aid in keeping clips in rhythmic alignment.

Figure 7.17 The Quantization Menu (right) with a value of 2 Bars selected

Recording on Scene Launch

The Session View can also be configured so that recording begins on all armed tracks upon launching a given scene. To do this, you need to enable the **Start Recording on Scene Launch** option in the Ableton Live Preferences window. This option is located under the **Record/Warp/Launch** tab.

Recording from Session View into an Arrangement

Arrangement Record is often under-utilized in Ableton Live as a way of recording playback from the Session View. This powerful workflow allows you to perform a song by launching scenes and clips in the Session View, while simultaneously recording the result into the Arrangement View.

Using Arrangement Record

Working with Arrangement Record is simple and intuitive, allowing you to focus on creatively performing the elements you've assembled in the Session View. Afterward, you can fine-tune the arrangement using the editing techniques you've learned about in the Arrangement View.

To record from the Session View to the Arrangement View:

1. Click the **Stop** button in the Control Bar twice to place the Arrangement Position at **1.1.1**—the very beginning of Bar 1. This will ensure that your recording begins at the start of the arrangement.

2. Click the **Arrangement Record** button (solid circle) in the Control Bar. The button will turn red, preparing the Live Set for recording into an arrangement. However, recording will not begin until you launch a clip or scene.

   Figure 7.18 The Arrangement Record button (red) in the Control Bar

3. Launch any combination of scenes and/or clips to perform your arrangement. Everything you do will be simultaneously recorded to the Arrangement View.

4. When the performance is finished, click the **Stop** button in the Control Bar or press the **Spacebar** to end the recording.

Playing Back an Arrangement Recording

Having completed the recording, it's time to switch to the Arrangement View and listen to your results.

Verify your recording in the Arrangement View:

1. Press **TAB** to switch from the Session View to the Arrangement View.

2. Click the red **BACK TO ARRANGEMENT** button at the top-right corner of the Arrangement display. This will cause Live to play back from the arrangement, rather than from the Session View clips.

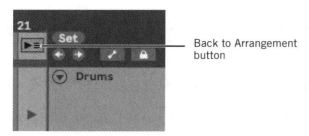

Figure 7.19 The Back to Arrangement button in the Arrangement View

3. Press the **STOP** button in the Control Bar twice (or press **HOME** on the computer keyboard) to position the cursor at the start of the Live Set.

4. Press the **PLAY** button in the Control Bar or press **SPACEBAR** to begin playback.

If you are not satisfied with your recorded arrangement, you can use **EDIT > UNDO** to undo the recording, or you can delete the clips. Then repeat the recording from the Session View as described above.

Review/Discussion Questions

1. Which Session View interface sections are not available in the Arrangement View? (See "The Session View" beginning on page 172.)

2. How is a clip's signal routed once it is launched in the Session View? (See "Clip Slots Section" beginning on page 173.)

3. What is the purpose of the Scenes section of the Session View? (See "Scenes Section" beginning on page 173.)

4. Which sections of the Session View are typically used for mixing tasks? (See "Session View Mixing Sections" beginning on page 175.)

5. How many Sends are available in Ableton Live Standard and Suite editions? How many sends are included in the default Ableton Live Set? (See "Send Section" beginning on page 176.)

6. When making an audio recording, will moving the slider in the Mixer Section of the Session View affect the input gain of the signal being recorded? (See "Mixer Section" beginning on page 176.)

7. What are some ways that clips can be imported to tracks in the Session View? How can you control what tracks or slots are used when importing multiple clips simultaneously? (See "Adding Clips to Tracks in Session View" beginning on page 178.)

8. How can you start and stop playback for a single clip in the Session View? (See "Starting and Stopping Playback" beginning on page 180.)

9. How can you start playback for an entire horizontal row of clips in the Session View? (See "Controlling Playback with Scenes" beginning on page 181.)

10. How can you change the project tempo and/or time signature using Scenes? (See "Setting Tempo and Time Signature Using Scenes" beginning on page 181.)

11. Why might you want to enable clip looping when working with clips in the Session View? How can you toggle clip looping on/off? (See "Enabling/Disabling Clip Looping" beginning on page 182.)

12. What process can you use to record directly to a Clip Slot in the Session View? (See "Recording Audio in Session View" beginning on page 183.)

13. How does the Record Quantization value impact the length of recorded audio or MIDI clips? (See "Session View Recording Quantization" beginning on page 184.)

14. What is the purpose of the Arrangement Record feature? (See "Using Arrangement Record" beginning on page 185.)

15. How can you switch Ableton Live's playback from the Session View to the Arrangement View? (See "Playing Back an Arrangement Recording" beginning on page 185.)

 To review additional material from this chapter and prepare for certification, see the Ableton Live Audio Production Basics Study Guide module available through the Elements|ED online learning platform at ElementsED.com.

Exercise 7

Working in the Session View

🎧 Activity

In this exercise, you will configure the Session View display. You will practice launching clips and scenes to mock up a song arrangement. Then you'll perform your song arrangement and record it into the Arrangement View.

⏱ Duration

This exercise should take approximately 15 minutes to complete.

⊕ Goals/Targets

- Play clips and scenes
- Loop-enable a clip
- Create a new scene
- Use Arrangement Record

Exercise Media

This exercise uses media files taken from the song, "Lights," provided courtesy of Bay Area band Fotograf.

Written by: Zack Vieira and Eric Kuehnl; Performed by: Fotograf

The media provided for this course may be used for educational purposes only. No rights are granted to use the media for any other personal, commercial, or non-commercial purposes.

Getting Started

To get started, you will launch Ableton Live and open a pre-built Ableton Live Set. If you are running Ableton Live Lite, you can use an alternate version of this Set designed for a lower track count.

Launch Ableton Live and open the Set file:

1. Launch Ableton Live. A new Ableton Live Set will be created and will display on screen.
2. Choose **FILE > OPEN LIVE SET**. The Open Document dialog box will appear.
3. Navigate to the location where you saved the Live APB Media Files folder in Exercise 1.
4. Locate the 07-Lights SV folder within the Live APB Media Files folder, and open the Lights-SV.als Set inside the folder. (If you are running Live Lite, open the Lights-SVLite.als Set file instead.) The Live Set will open.

> ⓘ If one or more samples are missing (offline), the Missing Files warning will appear in red at the bottom of the Ableton Live window. Click the warning message to resolve the issue.

See Appendix A for details on locating missing files.

5. Choose **FILE > SAVE LIVE SET AS**. The Save Live Set As dialog box will appear.
6. Name your Set Lights02-xxx, where *xxx* is your initials, and navigate to an appropriate save location on your system.

> ⓘ In a classroom or lab environment, you may be required to save or submit your work using a specific location. Check with your instructor for details.

7. Click the **SAVE** button at the bottom of the dialog box to save the Set.

Playing Clips and Scenes

Now that you've got the Set open, you'll see a variety of tracks and clips in the Session View. (See Figure 7.20.)

Working in the Session View

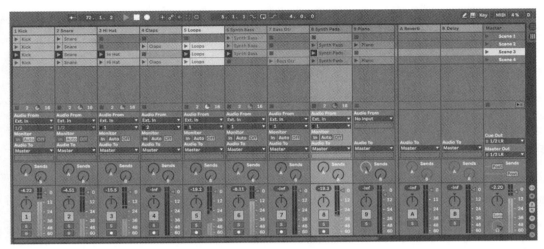

Figure 7.20 The exercise Set showing the Session View

Note that the tracks have already been color-coded:

- Drums/Handclaps: Orange
- Loops: Yellow
- Basses: Green
- Keyboards: Blue

The clips on the tracks have been organized into four scenes. Take a few minutes to launch each scene as well as individual clips to get a feel for the song.

To launch a scene, do one of the following:

- Click on the triangular **SCENE LAUNCH** button for the row on the Master track.
- Select the scene by clicking on it on the Master track; then press the **RETURN** key (Mac) or **ENTER** key (Windows).

To launch an individual clip:

- Click the triangular **CLIP LAUNCH** button at the left edge of a clip slot on a track.

To stop playback, do one of the following:

- Click the **CLIP STOP** button on any empty Clip Slot in the track (or below all the clip slots in the Track Status Display area) to stop that track from playing.

192 Exercise 7

- Click the **STOP ALL CLIPS** button on the Master track or the **STOP** button in the Control Bar, or press the **SPACEBAR**, to stop all tracks from playing.

Next you will rename the scenes to give them more descriptive names.

Rename the Scenes:

- Select each scene in turn and press **COMMAND+R** (Mac) or **CTRL+R** (Windows) to rename them each, as follows:
  - Scene 1 = Verse 1
  - Scene 2 = Interlude
  - Scene 3 = Verse 2
  - Scene 4 = Chorus

 After pressing COMMAND+R (Mac) or CTRL+R (Windows) and typing a scene name for the first scene, you can simply press TAB to move to successive scenes.

Loop-Enabling the Synth Bass Clip

When you launch the first scene, the Kick and Snare clips loop indefinitely, but the Synth Bass clip plays only once and then stops. This is because the Synth Bass clip is not loop-enabled.

Loop-enable the Synth Bass clip:

1. Double-click on the **Synth Bass** clip slot in the first Scene to display the clip in the Detail View at the bottom of the window.

2. In the Detail View's Sample box, click on the **LOOP** switch to loop-enable the clip.

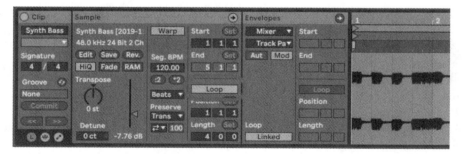

Figure 7.21 The Detail View with the Loop switch circled in red (shown enabled)

3. Play the first Scene again to verify that the **Synth Bass** clip will now loop indefinitely.

Creating an Introduction

Now that you've loop-enabled the **Synth Bass** clip, you'll use that clip to make an introductory scene. (This will sound similar to the bass-only introduction you created in the Arrangement View in Exercise 6.)

Create a new Scene:

1. Choose **CREATE > INSERT SCENE** or press **COMMAND+I** (Mac) or **CTRL+I** (Windows).

   ⓘ The Session View must have the active focus in order to create a new scene. If needed, click on the first Synth Bass clip slot or the Verse 1 scene to return the focus to the Session View prior to choosing Insert Scene.

2. Select the new scene name and press **COMMAND+R** (Mac)/**CTRL+R** (Windows) to rename it as **Intro**.

3. Drag the scene to the top of the Scenes section on the Master track to make it the first scene.

Now you're ready to place the **Synth Bass** clip into the new scene. The easiest way to do this is to drag a copy of the clip from another scene.

Drag and drop a copy of the clip:

- Hold down **OPTION** (Mac) or **ALT** (Windows) while dragging the **Synth Bass** clip from the **Verse 1** scene onto the **Intro** scene. A copy of the clip will be placed in the clip slot on the **Intro** scene.

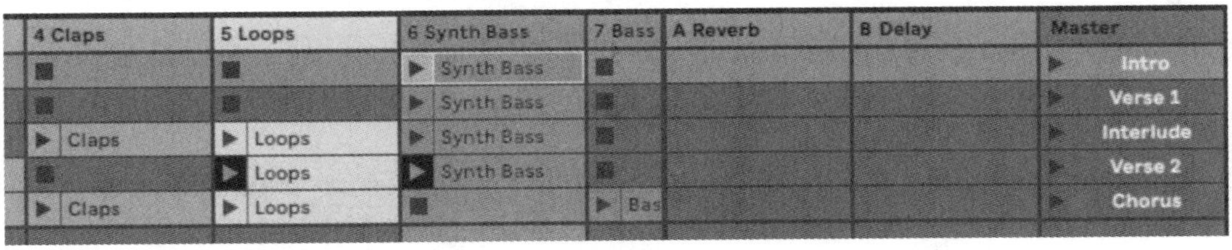

Figure 7.22 The new Intro scene including a copy of the Synth Bass clip

At this point, you should take a little more time to play through the scenes and figure out the final arrangement for a short version of this song (try for a duration of around 1:30).

- Practice playing through the scenes in succession, allowing each scene to play for one or more complete iterations.
- Consider returning to the **Intro** or **Verse 1** scenes after the **Chorus** scene, to serve as an ending.

Note that the **Interlude** and **Chorus** scenes are each 8-bars long, while the other scenes are all 4-bars long. Take this into consideration as you trigger successive scenes. Be sure to allow enough time for each scene to complete before starting the next scene.

Using Arrangement Record

Once you've determined how you would like to arrange the song, it's time to perform the arrangement in the Session View and record it over to the Arrangement View.

Record the Session View performance using Arrangement Record:

1. Click the **Stop** button (solid square) in the Control Bar twice to stop the transport and return the Insert Marker to the start of the Arrangement (1.1.1).

2. Click the **Arrangement Record** button (solid circle) in the Control Bar. The Arrangement Record button will turn red and the Live Set will be primed for recording.

Figure 7.23 The Arrangement Record button (red) in the Control Bar

> If Ableton Live goes directly into record when you click the Arrangement Record button, the Start Playback with Record preference is enabled. In this case, simply hold **Shift** when clicking Arrangement Record to temporarily toggle this behavior.

3. Click the **Scene Launch** button for the scene that you want to use at the beginning of the arrangement. This will start playback of the song and initiate the Arrangement Record pass.

4. Launch other scenes (or individual clips) to introduce them in the desired order and perform your arrangement.

> You must wait until the currently queued scene launches before clicking on a subsequent scene. You cannot queue up multiple successive scenes at once.

5. When finished, click the **Stop** button in the Control Bar to end the recording.

6. Press **Tab** to switch to the Arrangement View. Your complete arrangement will display exactly as you performed it. However, the clips will appear dimmed, indicating that the Arrangement View is inactive.

7. Click the **Back to Arrangement** button above the tracks to switch the project playback to the recorded arrangement. The clips will change to display as active. (See Figures 7.24 and 7.25.)

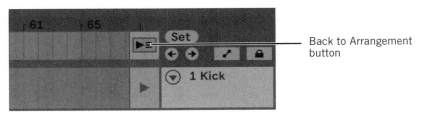

Figure 7.24 The Back to Arrangement button in the Arrangement View

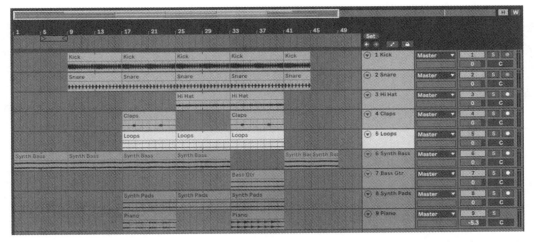

Figure 7.25 An example of a an arrangement created using Arrangement Record, with clips shown active

8. Play through the arrangement to verify that it sounds as expected.

Finishing Up

Feel free to experiment further with the project and edit the clips in the Arrangement View. When finished, remember to save the work you've done.

Save your work:

1. Choose **FILE > SAVE LIVE SET** to save the Set.

2. If desired, you can exit Ableton Live by doing one of the following:

   - On a Mac-based system, choose **LIVE > QUIT LIVE**.
   - On a Windows-based system, choose **FILE > EXIT**.

   Ableton Live will automatically close.

That completes this exercise.

Chapter 8

Mixing Concepts

...*What You Need to Know to Mix a Project*...

This chapter introduces you to mixing concepts, from setting levels and panning to the role of EQ and dynamics processing on the tracks in a mix. We discuss how to keep a mix from clipping, the difference between gain-based processing and time-based processing, and the differences between device processing and send-and-return processing. The chapter ends with a discussion of mixing in the box and some suggestions for maximizing your results when mixing in Ableton Live.

◉ Learning Targets for This Chapter

- Set effective levels for your tracks
- Recognize the role of EQ and dynamics processing in creating a balanced mix
- Use panning to create a sense of space and positioning in your mix
- Understand how devices and sends can be used to add processing to tracks
- Recognize the advantages of in-the-box mixing

Key topics from this chapter are illustrated in the Ableton Live Audio Production Basics Study Guide module available through the Elements|ED online learning platform. Sign up for a free account at ElementsED.com.

After you have recorded, imported, or otherwise created the media that you want to use in your project, you can set about the task of mixing the audio. Mixing is the process of setting the basic levels, positioning, and sonic characteristics of your tracks. Essentially, this is where you determine how the various parts of the project will blend together to create a cohesive result during playback.

Basic Mixing

Although mixes can get very complex on large sessions, the basic process is fairly simple. The primary goal when creating a stereo mix is to set the levels for each of the tracks using the tracks' Volume controls and to position each of the tracks within the stereo field using the tracks' Pan control. Other options include using devices and sends to process each track and add effects.

Setting Levels

Setting the levels for your tracks is typically done in the Session View. Here you'll find Volume controls for each track, which you can use to adjust each track's overall output level.

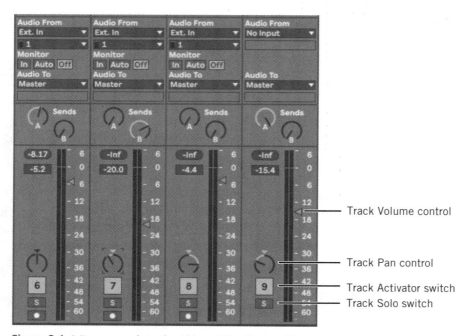

Figure 8.1 Mixer controls in the Ableton Live Session View

Level Considerations

The purpose of adjusting the output level of a track is to make the track audible without obscuring other tracks or causing the track to become overly prominent in the mix. During the recording process, audio is often recorded louder than it needs to be in the final mix. Additionally, as you begin summing multiple tracks together, the overall output level for the project will increase. For these reasons, it helps to pull all of the Volume controls down quite a bit when you begin mixing and then gradually adjust the levels of each track as needed.

Allowing one track to be heard above all the others is not only a matter of raising its Volume control, however. A common mixing dilemma is that making one track louder will cause another, equally important track to become overshadowed and indistinct.

> **Getting a Good Music Mix Is More Than Setting Levels**
> Raising the Volume control on a guitar track enough to get it to cut through the mix can make the vocal track hard to hear. Raising the Volume control on the vocal track may then begin to obscure the impact of the drums. And raising the Volume controls on the drum tracks can affect the bass guitar, which may now be barely audible. By the time you bring up the level on the bass guitar to compensate, you may be right back where you started, with the guitar track no longer cutting through the mix. Only now, everything is louder!

In the quest for the ultimate mix, where "everything is louder than everything else," often you simply end up with a very loud mix and still cannot distinguish individual parts.

As you start setting the initial levels for your mix, try to achieve a good overall balance. Don't get overly concerned if you find that certain tracks start to compete with one another. These issues can be addressed later using panning, EQ and dynamics processing, and other techniques that help give each track its own unique space in the mix.

Beware of Clipping

Among the problems caused by loud tracks, the output levels of your mix may get too "hot," leading the signal to clip at the digital-to-analog converters. This will cause the mix to become distorted on playback during the loudest moments.

Digital distortion is always detrimental to sound quality and should be avoided at all costs. Therefore, it is better to mix too quiet than it is to mix too loud. You can always increase the overall output levels of a mixed stereo file at a later stage if needed. But you cannot remove clipping from the stereo file after the fact.

Metering on Source Tracks

To help you keep an eye on your audio levels, each track includes a standard meter display to the left of the track Volume control. At the top of each meter is a clip indicator for the track. The clip indicator will light whenever the signal from the track exceeds full-scale audio.

Meters in the Ableton Live mixer measure loudness in decibels relative to full-scale audio (dBFS), with full scale represented by 0 at the top of the meter. Digital clipping occurs whenever a signal exceeds 0 dBFS at an input or output.

Figure 8.2 Meters on tracks in Ableton Live

Audio tracks show *post-fader* metering by default. This means that the meter reflects the output level of the track rather than the level of the source audio on the track (disk file). A lit clip indicator on an Audio track means the output levels are too hot—the Volume control should be lowered, to reduce the amplitude.

Metering on Master Tracks

In order to ensure that the output level of your Set is not too hot, it's a good idea to reference the metering on the Master track of your project. The level meter on a Master track reflects the overall output of the entire mix. This is a summed total of the signals from each contributing track.

Clip indicators on a Master track will light red when the summed signal level exceeds the capabilities of the digital-to-analog converters. In this case, you can reduce the levels on each of the contributing tracks in your mix, or reduce the overall output level using the Master track itself.

While working on your mix, keep an eye on the meters on the Master track. If the Master track begins clipping, reduce the output levels before continuing.

> ### Clips Versus Overs
>
> In Ableton Live, the meters on Audio tracks, MIDI tracks, and Return tracks do not change color to indicate an overage on the track (a signal exceeding 0 dBFS en route to an internal destination). This is distinguished from the meters on Master tracks, which will turn completely red to indicate a true clip (a signal that will distort an output or disk file, if not attenuated).

The Role of EQ and Dynamics Processing

As mentioned above, setting volume levels is not the only consideration when it comes to allowing each track to be heard in a mix. Other important considerations include the track's frequency spectrum and the track's dynamic response.

The frequency spectrum of a track, or the amount of energy the track has at different audio frequencies, can be shaped using equalization (or EQ for short). You can use EQ processing to help reduce low-end rumble, high-end hiss, and other unwanted noise in a signal. You can also use EQ to shape a signal and create a unique tone for each track. This allows certain target frequencies or frequency ranges to be more prominent than others. By selectively boosting and cutting frequencies on each track, you can balance the audio spectrum in the mix and give each track its own sonic footprint.

In the interest of keeping track levels in check, it is generally better to focus on cutting the frequencies you don't want rather than boosting the frequencies you wish to emphasize.

The dynamic response of a track refers to the range of amplitude values on the track. Said another way, the track's dynamics relate to the differences in loudness from one part of the track to another. The momentary loudness peaks (or *transients*) on a track are often much louder than the track's average levels. These loudness peaks can begin to obscure other tracks (or cause clipping) long before the source track's overall amplitude is where you want it in your mix.

A solution to this problem is to use compression on the track. Compression helps reduce the loudness in peak areas without affecting the average loudness. Using compression in combination with an appropriate amount of make-up gain can help you increase the level on a track without obliterating the other tracks in your mix.

For details on using EQ and dynamics processing, see the associated discussions in Chapter 9 of this book.

How to Set Levels

When it comes to setting the relative levels of your tracks, it helps to get familiar with the audio characteristics of each track first. Then consider how those characteristics contribute to the overall mix. Use the following steps as a guide to get started:

- **Listen to the track in isolation**—Use the Solo function to isolate each track in turn and familiarize yourself with its sonic characteristics. When working on a music mix, consider how the track supports the rhythm, groove, harmony, and melody of the piece. When working with non-musical material, consider how the track contributes to the tone and emotion of the mix, and listen for the clarity of the track.

- **Listen to the track in context**—Unsolo the track to determine the appropriate level for the track relative to the other tracks in the mix. Can you still hear the important sonic elements that you heard when the track was soloed? Or is the track completely lost among competing sounds?

- **Listen to the mix without the track**—Click the Track Activator switch to deactivate (mute) the track. Gauge the track's contribution to the mix and how necessary it is. Does the mix sound lacking without the track? Or is the mix suddenly clearer without the track competing for sonic space?

- **Check the track throughout the piece**—Toggle the Activator and Solo switches on/off for the track to hear the track isolated, in context, and removed from the mix in multiple locations throughout the composition. Are there some parts of the mix where the track is more important than others? Should the levels change for different sections of the mix?

- **Make adjustments in iterations**—Return to the track and adjust the levels as you make other changes to the mix. Setting levels is an iterative process that will require fine-tuning as your mix begins to take shape. Be sure to consider how each track's contribution changes as you begin setting levels on other tracks, adjusting pan positions, and adding processing to the mix.

Work your way though each track in the project, listening and adjusting levels as you go to create a good overall balance. Don't worry if there are aspects of the mix you aren't completely happy with at this point; you will be refining the result as you progress through later stages of the mixing process.

Panning

The pan controls on each track allow you to position the track within the stereo field.

Pan Controls in Stereo Pan Mode

Audio, MIDI, and Return tracks default to a single Pan control, operating in Stereo Pan mode. The default Pan control allows you to position the track's output as desired in the stereo spectrum. Panning a track hard left will cause it to play out of the left speaker only; panning to center will cause the track to play at equal volume out of both speakers; and panning hard right will cause the track to play out of the right speaker only.

Stereo Pan mode does not provide independent channel control when a track contains a stereo signal, however. Instead, the Pan control functions as a balance control in this mode. This means that panning the track hard left simply turns down the volume of the right channel and vice versa.

Figure 8.3 Tracks panned hard left, center, and hard right (from left to right)

Split Stereo Mode

To achieve true stereo panning when a track contains a stereo signal, you can switch the track to Split Stereo Mode. This provides greater flexibility in panning each channel.

To enable Split Stereo Mode on a track:

- **RIGHT-CLICK** on a Pan control and choose **SELECT SPLIT STEREO PAN MODE** from the context menu. The Pan control knob will be replaced by separate Left and Right pan sliders.

In Split Stereo Mode, the Pan function uses separate pan sliders to enable discrete positioning of each channel in the stereo field. Panning both channels to the same position will create a summed mono signal. This allows you to position the entire signal at a specific point within the stereo field, just as with a mono track.

Leaving the pan controls set hard left (for the left channel) and hard right (for the right channel) will preserve the stereo separation between the channels in the source audio. Panning each control partially towards the center will collapse the stereo spread, reducing the panoramic width of the track.

You can also pan each channel to its opposite extreme, effectively flipping the stereo image and reversing the signal's placement in the mix.

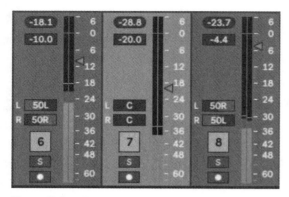

Figure 8.4 Split Stereo pan controls showing default panning, mono panning, and inverted panning (left to right)

Panning Examples

As you refine your mix, you'll want to consider the panoramic position of each track, along with the overall distribution of the tracks. Positioning individual tracks at distinct locations will give each track its own space in the mix and help give your mix width and realism.

- **Center–panned tracks**—Certain tracks tend to work best in the center of a mix. These typically include tracks that provide the main focus or lead part for the piece, such as a narration or voiceover track or a lead vocal track. In music production, it is also common to place the main rhythmic elements at or near the center of the mix, such as the kick drum, snare drum, and bass guitar.

- **Hard–panned tracks**—Tracks that provide flavor, color, or character to a mix are often effective when panned hard left or right. These are typically supplemental tracks such as a shaker or harmonica in a music mix or ambient accents such as a distant siren or muffled argument from an adjacent room in a fiction narrative.

 Hard panning can also be effective on main tracks, especially in a sparse mix.

One potential problem with using hard-panned tracks is that they can become inaudible to a listener who is on the opposite side of the listening space from the sound source, such as when the stereo speakers are placed wide apart in a living room. Also, if the stereo playback should drop a channel, everything hard-panned to the dropped side would be lost.

These problems can be mitigated by not panning fully left/right, by including some reverb from a hard-panned track in the opposite channel, and by using psycho-acoustic processing to place an image in a stereo field by means of timing offsets while keeping the signal in both channels.

For an example of hard panning, check out the electric guitar on early albums by Van Halen. Hard panning is prominent in songs such as "Little Dreamer," "Beautiful Girls," "Hang 'Em High," and numerous others.

- **Tracks panned off-center**—Often tracks are placed off center to create realism and a sense of space in the mix. This technique is commonly used to position characters around a room in a dialog mix. In music mixing, it is common to pan individual drum tracks to emulate the layout of the drum kit (placing hi-hat and splash cymbal on the left, rack toms across the middle, and floor tom and ride cymbal on the right, for example). Another common technique is to pan backing vocal tracks, placing high harmonies on the left, mid-range parts near the center, and low harmonies on the right, for example.

- **Other ideas**— It can be very effective to offset parts that have similar impact in a mix. For example, panning a doubled vocal part to left and right extremes can create a sense of width in the mix. Similarly, panning a rhythm guitar part opposite a Rhodes piano part can help each part contribute equally without competing with one another.

 For examples of effective panning for multiple vocal parts, take a listen to "Ziggy Stardust," "Changes," and other classics by David Bowie. Bowie often used opposite-panned vocal doubles and harmonies during key sections for emphasis.

Processing Options and Techniques

When it comes to adding processing to your tracks, you have two broad categories to consider: gain-based processing and time-based processing. You also need to consider when and how any processing should be added to your mix.

Gain-Based Processing

Gain-based processors include any processors that affect the amplitude of the audio signal in some way. Examples include EQs, compressors, noise gates, expanders, and similar dynamics processors.

When adding gain-based processing to a signal, you will typically assign the processor as a device on an individual track. This type of processing is usually applied to the entire signal (100% wet), rather than being mixed together with a dry signal. The processing is also commonly track-specific.

For example, when using an EQ plug-in to eliminate low-frequency rumble, you will want 100% of the source signal to be affected. And you will adjust the EQ parameters to address the specific frequency characteristics of the source signal on the track.

Time-Based Processing and Effects

The other type of processing can broadly be classified as time-based processing and effects. This includes processors that affect the signal in the time-domain, such as reverbs and delays, and modulation effects, such as choruses, flangers, and phasers. These processors typically get applied to only a portion of the signal and mixed back in with the dry signal.

Time-based processors are commonly used as shared effects, across multiple tracks in a mix. This helps provide consistency in the mix and makes more efficient use of your processing resources.

Inserted Devices Versus Sends

Ableton Live allows you to add signal processing to a track using devices inserted directly on the track and/or using sends. A device is added to a track using an audio patch point that places the processor directly into the signal path of the track. When using a device, all of the audio on the track must pass through the device on its way to the track's output.

Ableton Live provides unlimited devices on each track, allowing you to process the track's signal through multiple successive devices, as needed.

 See Chapter 8 in this book for details on using devices for signal processing.

By contrast, a send provides a signal path that can be used to route audio from one or more source tracks to a parallel destination for processing. In Ableton Live, signals from sends are returned to the mix by way of a Return track. Processing is commonly added to the signal using a device on the Return track.

When you create a Return track for internal device processing, Ableton Live will automatically route Send signals from all tracks to the destination Return track.

Processing with External Gear

In Ableton Live, you can also choose to route the signal from a track through external hardware, provided you have sufficient I/O on your audio interface. In this case, the signal will route out of your audio interface, through an external processor, and back into your audio interface, before the signal is returned to the mix. (See Figure 8.5.)

Figure 8.5 External reverb connected to an audio interface for use with Ableton Live

Mixing in the Box

Although it is possible to use external gear for processing in Ableton Live, mixing completely "in the box" offers numerous advantages. Mixing in the box simply means that all signal processing is provided by internal software devices, so the mix does not rely on any external gear.

Advantages of In-the-Box Mixing

The advantages of working completely in the box include the following considerations:

- **Portability**—Creating your Ableton Live mix entirely in the box means that you can open the project from any Ableton Live workstation anywhere in the world, as long as you have an Internet connection. Barring any missing devices, the mix will sound the same on any system.

  If you were to use external gear, you will need to take the gear with you to work on the mix from a different location.

- **Recallability**—Saving the project will save all device settings and levels, allowing them to recall exactly as they were the next time the project is used. This is not the case when using external gear, as you will need to reconfigure all settings on the external hardware to match your last used settings for the project.

- **Time savings**—Reconfiguring hardware is not only inconvenient, it can be quite time-consuming. It also requires keeping detailed notes of settings and configurations, and updating those notes every time you change a setting.

- **Dynamic automation**—Device settings can be automated, enabling you to incorporate dynamic changes in your mix. All automation settings are saved and recalled with the project, ensuring that it plays back consistently every time.

Getting the Most Out of an In-the-Box Mix

To get the most out of an in-the-box mix, you will want to consider how much control you need to have over the sonic characteristics of your mix. Ideally, this is something you would begin thinking about at the project inception, prior to starting to record.

For example, you'll need to decide early on if you want to be able to control the dynamics of each track entirely from within the project, or if it is OK to apply some compression to individual signals prior to the recording input stage. Likewise, you'll need to decide whether to record sources such as guitar tracks from an amp, with the guitarist's effects processing already in place, or to record a direct signal from the guitar and add amp simulators and stomp-box effects as devices inside of Ableton Live.

Aside from considerations about when to apply processing, you will also need to take steps to maximize the results of any processing you apply from within Ableton Live. Use the following simple suggestions as a starting point:

- **Learn how to set device parameters**—To effectively use the devices that are provided with Ableton Live, you will need to know how to set the parameters properly and recognize what each parameter is used for. Take time to experiment and practice using each device on material you are familiar with.

- **Avoid using copies of the same device on many different tracks**—Learn how to use send-and-return processing to apply the same effects to multiple tracks, rather than placing individual copies of the same device on each track. Not only does using multiple copies of a device require more processing power than using a single copy, but you'll find that keeping the settings in sync across multiple tracks can become tedious and time-consuming.

- **Invest in good quality plug-ins**—The device collection that comes with Ableton Live is sufficient to get you started. As you expand beyond that basic set, invest in quality processors that enhance your sonic palette. Don't be afraid to spend some money to get what you want, but keep in mind that many high-quality plug-ins are available at very reasonable prices. Spend some time reading or viewing product reviews, and check out trial versions, if available. Once you know what you want, keep an eye out for discounts and sale prices.

> ⓘ Plug-in manufacturers commonly offer discounts and promotional pricing during national holidays, trade shows, and similar events.

Review/Discussion Questions

1. What are some reasons for lowering the Volume controls across your entire project as you begin to set levels for your mix? (See "Level Considerations" beginning on page 198.)

2. Why is it important to avoid clipping in your mix? How can you tell if the project is clipping at the outputs? (See "Beware of Clipping" beginning on page 199.)

3. How can equalization be used to help a track cut through a mix? (See "The Role of EQ and Dynamics Processing" beginning on page 201.)

4. What is meant by the dynamic response of a track? What can be done to "tame" excessively loud peaks on a track? (See "The Role of EQ and Dynamics Processing" beginning on page 201.)

5. What two pan modes are available in Ableton Live and how are they different? (See "Panning" beginning on page 202.)

6. Describe some examples of panning techniques that can be useful or effective in a mix. (See "Panning Examples" beginning on page 204.)

7. How does gain-based processing affect an audio signal? Give some examples of gain-based processors. (See "Gain-Based Processing" beginning on page 205.)

8. What are some available time-based processors? How are time-based processors typically applied to tracks? How is this different from the way gain-based processors are used? (See "Time-Based Processing and Effects" beginning on page 206.)

9. How are inserted devices different from sends, in the way they process audio? What is the role of a Return track in a send-and-return configuration? (See "Inserted Devices Versus Sends" beginning on page 206.)

10. What is meant by "in-the-box" mixing? What are some advantages of mixing in the box? (See "Mixing in the Box" beginning on page 207.)

11. Why is it important to begin thinking about in-the-box mixing from the project inception? (See "Getting the Most Out of an In-the-Box Mix" beginning on page 208.)

12. How can you maximize the results of any processing you apply from within Ableton Live? (See "Getting the Most Out of an In-the-Box Mix" beginning on page 208.)

 To review additional material from this chapter and prepare for certification, see the Ableton Live Audio Production Basics Study Guide module available through the Elements|ED online learning platform at ElementsED.com.

Exercise 8

Creating a Basic Mix

Activity

In this exercise, you will perform basic mixing tasks in Ableton Live. You'll start off by adjusting levels to get a good starting mix. Then you'll explore some strategies to prevent clipping for the Set. Finally, you'll adjust pan settings to enhance the mix and create space for the different instruments.

Duration

This exercise should take approximately 20 minutes to complete.

Goals/Targets

- Set levels for different tracks in the project
- Use the Master track to monitor peak levels and prevent clipping
- Set panning for different tracks in the project
- Save your work for use in Exercise 9

Exercise Media

This exercise uses media files taken from the song, "Lights," provided courtesy of Bay Area band Fotograf.

Written by: Zack Vieira and Eric Kuehnl; Performed by: Fotograf

The media provided for this course may be used for educational purposes only. No rights are granted to use the media for any other personal, commercial, or non-commercial purposes.

Getting Started

To get started, you will open the multi-track project you created in Exercise 6. This will serve as the starting point for this exercise.

Open your existing Lights project:

1. Launch Ableton Live, if it isn't already running. Once startup completes, the Default Set will be visible.
2. Select **FILE > OPEN LIVE SET**.
3. Navigate to the location where you saved your Lights01-xxx.als file in Exercise 6.
4. Open the Lights01-xxx project. The Set will open as it was when last saved.
5. Choose **FILE > SAVE LIVE SET AS** to save a copy of the Set under a new name.
6. In the resulting dialog box, name the copy Lights03-xxx, where xxx is your initials, and click **SAVE**.
7. Press the **TAB** key to toggle the display to the Session View.

Creating a Rough Mix

At this point, you're ready to start mixing. As you adjust the Volume controls on the tracks in your project, you will want to keep an eye on your output levels.

Create the Rough Mix:

1. Press the **SPACEBAR** to begin playback.
2. Listen to the levels of each track. Begin evaluating the changes you'll need to make to even out the parts. You want each part to be heard clearly without "jumping out" of the mix and sounding too prominent.
3. See the suggestions below to help you evaluate the tracks and begin making changes.

General Mixing Suggestions

Here are some general tips for setting the rough levels for tracks:

- Solo a track and listen closely to the content. What are the most important details of that instrument? When you un-solo the track, can you still hear those details clearly in the overall mix? If not, you may need to increase the level of the track.

- Deactivate a track and listen to the mix. Does the mix sound empty without that track? Do you even notice that it is missing? You may decide to leave the track deactivated or reduce the level of the track.

- While listening through the whole song, consider whether a track is necessary only in a particular section of the song. Most songs build in complexity as they progress. You might want to consider muting or reducing the volume of a track at the beginning of the song or during verses. Then you can unmute or increase the volume of the track later in the song or during choruses.

- Keep revisiting the various sections of the song. As you make changes in one section, it may change how you feel about other sections.

Suggestions for This Project

Here are some suggestions to help you create a rough mix of your project. As you work, press the **TAB** key to toggle to the Arrangement View to position the Insert Marker; then press **TAB** again to toggle to the Session View to adjust the Volume controls and other mixer settings for your tracks as needed.

- Listen to the mix going into Chorus 1 (around Bar 48). The **Synth Lead** and **Piano** tracks are a bit too loud in these sections. Try reducing the level of each track by about –5 dB or until the levels sound balanced during playback.

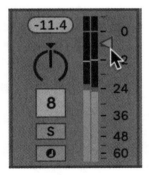

Figure 8.6 Lowering the Volume control to around –5 dB below the default 0 position

> If you are using the Set for Ableton Live Lite, you will not have a Synth Lead track. In this case, make the change to the Piano track only.

- Listen to the mix in Chorus 2 (around Bar 77). The **Strings** track is much too loud (not applicable in Ableton Live Lite Set). Try reducing the level on this track by around –8 dB.

- The peaks on the **Bass Gtr** track get a little too loud during both choruses (not applicable in the Ableton Live Lite Set). Try reducing the volume on this track by about –2 dB.

The Set for Ableton Live Lite does not have a separate Bass Gtr track. The level of the bass guitar has already been attenuated in the Basses track.

- Throughout the song, the vocal track (Vox) gets a little bit lost in the mix. Try increasing the level on this track by around +2 dB.

Adjusting Levels to Prevent Clipping

When mixing a project, it is fairly common to have some level overages on individual track meters or clipping on the Master track. This can be tricky to deal with when you're first developing your mixing skills. Fortunately, you have a number of ways to address clipping in Ableton Live.

General Level Suggestions

Here are some tips for dealing with clipping:

- If an individual track is showing level overages, the first thing to do is reduce that track's level. If you can prevent clipping while the resulting level still sounds good in the mix, you're all set. If the resulting level is too quiet, you'll need to try another approach.

- Another option is to reduce the levels of all of the tracks. To do this, first reduce the level of the track that is causing the clipping until the clip goes away. Take note of how much you reduced the track from its prior level in the mix (such as –6 dB). Then reduce the level of each track in the session by the same amount. Afterwards, you may need to bring up the level of the Master track to get back to a reasonably loud mix.

 You can adjust the relative levels of multiple tracks simultaneously but selecting the desired tracks and then moving the Volume control on any selected track.

- A third option to prevent clipping is to simply lower the Volume control on the Master track to correct the clip at the output stage. While overages on an individual track may not be detrimental to the final mix, you never want to have clipping on the Master track.

Suggestions for This Project

The individual tracks in this mix are well below clipping, so you should not need to reduce them further. But the overall mix is clipping. You'll use the Master track to solve that problem.

Correct clipping at the output stage:

1. Play through the song in its entirety.

2. Keep an eye on the **Master** track, watching the level on the meters and the Peak Level button at the top of the meters. (See Figure 8.6.)

3. Take note of the loudest areas of the project. After playing the song all the way through, you should have an idea of where the signal is peaking.

Figure 8.6 The Peak Level button on the Master track can be seen at the top left in yellow.

4. Reduce the level of the **Master** track by at least the amount shown in yellow in the Peak Level button. (Alternatively, if the peak is below zero, you could raise the level until the peak is just below clipping.)

> ⓘ Hold Shift when moving a track's Volume control to have fine control over the level.

5. Reset the Peak Level button by clicking on it to clear the currently displayed value.

6. Play through the loudest sections of the song again to verify that the peak levels now remain below clipping. If clipping still occurs, reduce the Master track level further and test again.

Adjusting Panning

In this part of the exercise, you'll adjust the panning for your tracks. Tracks in this project include both mono and stereo clips. Mono clips can be easily positioned to any location in the stereo field. Stereo clips are often adjusted to have a narrower or broader width. Both of these techniques can be used to add interest to the mix, and can also create space where each track can be heard and appreciated.

General Panning Suggestions

Here are some general tips for panning tracks:

- For tracks with mono clips, solo the track and adjust the panning. Listen for a particular pan position that just "feels right" for the track or instrument. Unsolo the track to check how its position works in the overall mix.

- For tracks with stereo clips, switch to Split Stereo pan mode to adjust the panning independently for the left and right channels. You may be able to create more space in the mix by narrowing some stereo tracks. See if any tracks sound better at the default (widest) pan setting instead.

- Try panning some tracks in unusual ways. Tracks with mono clips can be panned all the way left or right to create a novel sound. Tracks with stereo clips can be shifted so that they are not symmetrical in the left and right channels. Just be careful because extreme panning can also distract from the listening experience.

Suggestions for This Project

Here are a few thoughts for altering the panning of the current project:

- Try setting the **Hi Hat** track slightly to the left (around **15L** [the 10 o'clock position]) and the **Claps** track slightly to the right (around **15R** [the 2 o'clock position]). This can create a more interesting mix of the percussion elements. (You can hear the effect of these changes in the choruses, where both tracks are contributing to the mix.)

> (i) The Set for Ableton Live Lite does not have separate tracks for the Hi Hat and Claps. In this case, set the pan for the Hats & Claps track slightly left.

- Create some space for the vocal by narrowing the **Vox** track a bit.

  - Enable Split Stereo pan mode by **RIGHT-CLICKING** on the Pan control and choosing **SELECT SPLIT STEREO PAN MODE** from the context menu.

Figure 8.7 Selecting Split Stereo pan mode

  - Narrow the pan width by setting the left pan to around **35L** and the right pan to around **35R**.

- Try using Split Stereo pan mode on the **Bass Gtr** track as well. Narrow the stereo image of the track to help it stand out. Set the left pan to around **20L** and the right pan to around **20R**. (You can hear the contribution of this track during each of the choruses.)

> ⓘ In the Ableton Live Lite Set, try making this change on the Basses track instead. You may not want to narrow the image as much in this case.

- Use Split Stereo pan mode on the **Strings** track (not applicable in the Ableton Live Lite Set). Move the image to the left by setting the right pan slider to center (**C**). (You can hear the contribution of this track in chorus 2.)

Finishing Up

To complete this exercise, you will need to save your Set and close the project. You will be reusing this project in Exercise 9, so it's important to retain the work you've done.

Before you wrap up, you can also listen to the project to hear the mixing changes you've made.

Finish your work:

1. While the transport is already stopped, press the **STOP** button to position the Insert Marker at the beginning of the arrangement.

2. (Optional) Press the **SPACEBAR** to begin playback and listen through the project. Press the **SPACEBAR** a second time when finished to stop playback.

3. Choose **FILE > SAVE LIVE SET** to save your work.

4. If desired, you can exit Ableton Live by doing one of the following:

   - On a Mac-based system, choose **LIVE > QUIT LIVE**.
   - On a Windows-based system, choose **FILE > EXIT**.

   Ableton Live will automatically close.

That completes this exercise.

Chapter 9

Signal Processing

...*What You Can Do to Optimize Your Audio*...

This chapter begins with an overview of using plug-in devices in Ableton Live. We take a look at some general categories of plug-ins—including EQ, dynamics, and time-based effects—and provide some suggestions for using various plug-in processors and parameters. We also look at the process of creating send-and-return configurations and discuss how this setup can be useful.

⊕ Learning Targets for This Chapter

- Understand basic device functionality in Ableton Live
- Become familiar with the major plug-in categories
- Learn how to use basic device parameters
- Explore send-and-return configurations

Key topics from this chapter are illustrated in the Ableton Live Audio Production Basics Study Guide module available through the Elements|ED online learning platform. Sign up for a free account at ElementsED.com.

Devices (or "plug-ins," as they are commonly known outside of Ableton Live) are the key to one of the most important aspects of working with a DAW: processing audio and adding effects. Innovations in plug-in design were a key factor in the rise of the DAW as the dominant platform for producing music and creating audio for film, TV, and video games. Without great–sounding plug-ins for equalization, dynamics processing, and time-based effects such as reverb and delay, the DAW revolution might never have occurred.

In this chapter, we take a look at how to use devices and plug-ins in Ableton Live and review the most common types of processors used in today's production workflow.

Processor Basics

The sheer variety of devices and plug-ins available today can seem completely overwhelming when you're just getting started. In order to ease the learning curve, Ableton has selected a core set of devices and included them in Ableton Live. Depending on the edition of Ableton Live you are using, the selection can be limited (Live Intro) or quite huge (Live Suite)! Learning just a few processors well is a realistic goal for those who are new to the DAW world. And, because there are only a few major categories of device/plug-in processors, the knowledge that you gain now can easily be applied to other Live devices and VST- or AU-format plug-ins down the road as you expand your processing palette.

 Ableton Live allows you to insert an unlimited number of devices on a track. The only limit is the processing power of your computer.

Devices Versus Plug-Ins

The collection of devices that come with Ableton Live can be extended with third-party plug-ins. Live supports Steinberg's VST plug-in format, as well as the AU plug-in format for Mac-based systems. Working with plug-ins in Live is very much like working with Live's native devices, with the exception that plug-ins are listed separately in the Live Browser, under the **Plug-Ins** category.

Note that plug-ins will not appear in the **Plug-Ins** category until you activate your plug-in sources. This tells Live which plug-ins you want to use and where they are located on your computer. (See the Ableton Live User Manual for details.)

For purposes of the discussion in this chapter, we use the terms "plug-in" and "device" interchangeably to refer to both the device processors that come with Ableton Live and compatible third-party plug-in processors, unless otherwise noted.

Viewing Devices on Tracks

Before you can place a device on a track, you'll need to know how to display Ableton Live's Device View. The Device View is actually one of the available modes of the Detail View. If the Detail View is currently hidden, there are several ways to make it visible.

To show (or hide) the Detail View, do one of the following:

- Select VIEW > SHOW/HIDE DETAIL VIEW.
- Press OPTION+COMMAND+L (Mac) or CTRL+ALT+L (Windows).
- Click the SHOW/HIDE DETAIL VIEW selector at the bottom right of the main window.

Figure 9.1 The Device View

Once the Detail View is visible, you'll need to make sure that it is displaying the Device View and not the Clip View.

To show devices in the Detail View, do one of the following:

- Select VIEW > TOGGLE CLIP/DEVICE VIEW.
- Double-click on the name of a track.
- Press SHIFT+TAB on the computer keyboard.

> Note that all of the above techniques for displaying the Device View will automatically show the Detail View if it was previously hidden.

Inserting and Removing Devices

In earlier chapters we looked at ways to insert devices (including virtual instruments) on tracks in Ableton Live. Let's do a quick recap before moving on to using devices to process audio.

To insert a device on a track:

1. Using one of the techniques described above, make sure that the Device View is visible.

2. Locate the desired device in the Browser.

3. Drag and drop the device into the Device View for the track. The device's interface will automatically be displayed.

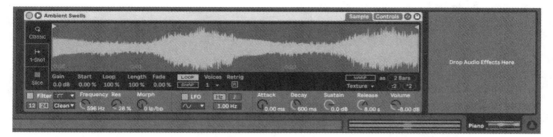

Figure 9.2 The Sampler instrument visible in the Device View

You can add additional devices to the track by dragging them before or after existing devices in the Device View. Device processing is applied to the track signal sequentially, from left to right.

 Devices can be collapsed (folded) in the Device View by double-clicking on the Device Title Bars. This can make it easier to see and arrange multiple devices on a track at once.

To remove a device from a track:

1. Using one of the techniques described above, make sure that the Device View is visible.

2. Click the Device Title Bar of the device you would like to remove.

3. Press the **DELETE** key on the computer keyboard.

Moving and Duplicating Devices

After you've begun working with devices on tracks in a project, you may find that you need to move a device to a different position or to a different track entirely. You may also want to create a duplicate of the device elsewhere in the project. These tasks are easily accomplished in Ableton Live.

To move a device, do one the following:

- Click on the Device Title Bar and drag it to the left or right of other devices on the same track.

- Click on the Device Title Bar and drag it onto a different track in the Session View or Arrangement view. The Device View will switch to show the devices on the target track.

To duplicate a device, do the following:

- Hold **OPTION** (Mac) or **ALT** (Windows) while clicking on the Device Title Bar and dragging the device to the desired location (on the same track or on another track).

Displaying Plug-In Windows

Before you can adjust the parameters of a third-party plug-in, you'll need to open the plug-in window. Plug-in windows open automatically when you assign a plug-in to a track. You can easily re-open a plug-in window that is not displayed at any time.

To open a plug-in window:

- Click on the **SHOW/HIDE PLUG-IN WINDOW** button (wrench icon) on the plug-in title bar in the Device View.

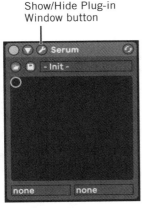

Figure 9.3 Clicking the Show/Hide Plug-In Window button for the Serum plug-in

 You can have Ableton Live automatically show and hide plug-in windows when you select a track by enabling the **AUTO-OPEN PLUG-IN WINDOWS** and **AUTO-HIDE PLUG-IN WINDOWS** in the Plug-Ins Preferences.

Common Device Controls

Ableton Live's devices feature two sets of controls: the standard controls that are common to all devices, and the process-specific controls that are unique to each device. Let's take a look at the common controls first. We'll look at process-specific controls later in this chapter.

Figure 9.4 The common controls shared by all devices in Ableton Live

Device Activator Button

On the left side of the Device Title Bar, you can see the Device Activator button (round yellow dot). You can click this button to activate or deactivate the device. When deactivated, the dot will appear grey. A deactivated device does not use any CPU processing power and does not modify the audio signal.

Hot-Swap Presets Button

On the right side of the Device Title bar, you can see the Hot-Swap Presets button (circle with rotation arrows). This button enables Hot-Swap mode for the device. When Hot-Swap mode is enabled, a temporary link is created between the device and the Browser allowing you to browse and select device presets.

To enable Hot-Swap mode, do one of the following:

- Click the **HOT-SWAP PRESETS** button in the Device Title bar.
- Press the **Q** key on the computer keyboard.

The button will change to display a red **X**.

 If a device is currently selected, pressing the Q key will put the selected device into Hot-Swap mode. If no device is selected, pressing the Q key will put the first audio effect (on audio tracks) or the first instrument (on MIDI tracks) into Hot-Swap mode.

Once a device is in Hot-Swap mode, you can use the computer keyboard to quickly browse and select device presets in the Browser.

To select presets in the Browser:

- Use the **UP/DOWN ARROW** keys to scroll through device presets.
- Use the **LEFT/RIGHT ARROW** keys to open and close preset folders.
- Press **ENTER** to load a selected preset.

To exit Hot-Swap mode, do one of the following:

- Click the **HOT-SWAP PRESETS** button in the Device Title Bar.

- Press the **Q** key on the computer keyboard.
- Press the **Esc** key.

Save Preset Button

To the right of the Hot-Swap Presets button, you can see the Save Preset button (circle with a floppy disk icon). Pressing the Save Preset button will store the current state of the device in the Browser's User Library.

To save a new preset:

1. Click the device's **Save Preset** button. The Browser will automatically switch to the User Library category and open the preset folder for the device.

2. Do one of the following:

   - To accept Ableton Live's suggested preset name, press **Return** (Mac) or **Enter** (Windows).
   - To create a new name for the preset, type the name and press **Return** (Mac) or **Enter** (Windows).

> ⓘ You can also save a new preset by dragging and dropping the Device Title Bar to the desired location in the Browser.

Adjusting Device Parameters

Once you have a device inserted on a track with the device visible in the Device View, you're ready to start adjusting the device's parameters. This is where the fun begins! Ableton Live provides a number of ways to adjust device parameters, including the obvious (using the mouse), and the not-so-obvious (using the computer keyboard). Let's take a closer look.

Adjusting Device Parameters with the Mouse

The most straightforward way to adjust device parameters is to use the mouse. For fader-style controls, simply click and drag up to increase the value of the control or drag down to decrease the value. To adjust knob-style plug-in parameter controls, click and drag vertically to rotate the control.

> ⓘ Unlike plug-ins developed by some other companies, Ableton's plug-ins do not respond to clicking at the location where you'd like a knob to go. You must click and drag up or down to change the knob's position.

You can change the way that device controls respond to the mouse using keyboard modifiers.

For fine adjustments:

- Hold **SHIFT** while moving a control.

To return a control to its default value:

- Double-click on the control.

Adjusting Device Parameters with the Computer Keyboard

It's actually a little-known fact that you can also use the computer alphanumeric keyboard (or "QWERTY" keyboard) to adjust device parameter controls in Ableton Live.

To change the value of plug-in parameter controls using the computer keyboard:

1. Click on the control that you'd like to adjust.
2. Do one of the following to change the control's value:
   - Press the **UP ARROW** to increase the value.
   - Press the **DOWN ARROW** to decrease the value.
   - Type in a numeric value and press **RETURN** (Mac) or **ENTER** (Windows).

EQ Processing

EQ (short for equalization) is probably the most commonly used processor in the audio universe. In the consumer electronics world, EQ can be found just about anywhere that an audio signal is present; even the most humble car stereo usually has basic low, mid, and high EQ controls.

In the realm of audio production, EQ is an absolutely essential tool for getting individual instruments to sound their best, and for getting a multi-track mix to sound great.

Types of EQ

In the simplest terms, EQ is used to manipulate the frequency component of an audio signal. However, EQ processors themselves come in a variety of flavors that offer different numbers of EQ bands, and different configurations within those bands.

Some types of EQ bands include:

- **Parametric**—This is the most common type of EQ band. A parametric band is fully adjustable in terms of the frequency, gain, and Q (width) parameters.
- **High- and Low-Pass Filters**—These constitute another common type of EQ band, and are often abbreviated HPF and LPF, respectively. These filters permit signal to "pass" only above (HPF) or

below (LPF) the selected frequency. High-pass filters are commonly used to eliminate low-frequency rumble, while low-pass filters can be employed to remove high-frequency hiss.

- **Shelf**—The shelf is similar to the high- and low-pass filter, but it is used to boost or cut the signal above (high-shelf) or below (low-shelf) the selected frequency by a specified amount.

- **Notch**—A notch filter is essentially a parametric band that has the gain setting fixed at the maximum negative value. This type of filter is used to "notch" out an offending frequency to make it inaudible.

Basic EQ Parameters

You'll find a number of standard EQ parameters on most software and hardware EQ processors. These will vary somewhat based on the type of EQ band.

Common EQ parameters include the following:

- **Frequency**—The Frequency parameter determines the target frequency for an EQ band. Every type of EQ band will have a frequency parameter, although some may be fixed at a specific frequency.

- **Gain**—The Gain parameter is used to boost or cut the volume of the associated EQ band. A gain control is typically found on parametric and shelf bands.

- **Q**—This is probably the most confusing EQ parameter for those who are new to audio production. The Q control is used to adjust the shape (slope or width) of an EQ curve. It behaves somewhat differently depending on the type of EQ band being used (parametric, shelf, filter).

- **In**—The In parameter can be used to enable or disable an EQ band, thus controlling whether that particular band is having an audible effect on the signal. This can be useful for doing an A/B comparison of the signal with and without an individual EQ band.

- **Input/Output** – The Input and Output gain controls can be used to boost or cut the signal at the input (before the EQ has been applied) or output (after the EQ has been applied).

EQ Devices in Ableton Live

Ableton Live 10.1 and above offers three types of EQ plug-ins: Channel EQ, EQ Three, and EQ Eight.

Channel EQ

The Channel EQ is a simple four-band EQ, featuring a high-pass filter, a low-shelf band, a semi-parametric mid-band, and a high-shelf band.

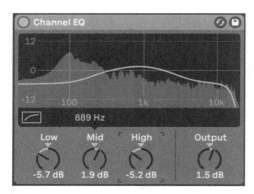

Figure 9.5 The Channel EQ device in Ableton Live

EQ Three

The EQ Three is a DJ-style EQ featuring low-, mid-, and high-bands with kill switches.

Figure 9.6 The EQ Three device in Ableton Live

EQ Eight (Ableton Live Standard and Suite Only)

The EQ Eight device is a flexible, high-quality 8-band EQ. Each of the eight bands in EQ Eight can be individually set to any of the following modes:

- Low Cut (high-pass filter)
- Low Shelf
- Bell (parametric)
- Notch
- High Shelf
- High Cut (low-pass filter)

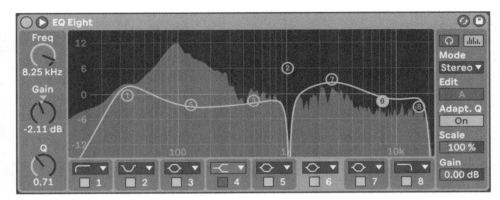

Figure 9.7 The EQ Eight device in Ableton Live

Strategies for Using EQ

Different types of material will require different approaches to EQ. But the basic techniques for finding good EQ settings are similar from track to track. Following are some tips for using EQ Eight and its bands.

To use an EQ Eight on a track:

- Enable a **LOW CUT** and increase the frequency enough to eliminate any unwanted noises at the extreme low frequencies, but not so much that the track loses too much low end.

    On some tracks, such as drum overheads, you may want to eliminate almost all of the bass to make room for the kick drum track and other low-frequency instruments.

- Use multiple **BELL** bands in the low-mid, mid, and/or high-mid frequency ranges to emphasize desirable qualities of a particular track.

    A common approach used to identify a target frequency is to boost the gain substantially (+12 dB or more) and slowly sweep the frequency control back and forth during playback. Listen carefully to the track while adjusting the frequency. Stop sweeping when you hear something you like, tone-wise. Then reduce the gain until the frequency is only slightly pronounced, compared to the original.

- Use multiple **BELL** bands in the low-mid, mid, and/or high-mid frequency ranges to reduce undesirable qualities of a particular track.

    A similar technique to the above can be used to identify target frequencies to cut. Simply stop sweeping when you hear something in the tone that sounds bad, intrusive, or unwanted. Then adjust the gain to a negative value that effectively reduces the prominence of the frequency, compared to the original.

- If necessary, use a **HIGH** or **LOW SHELF** to add some high-end sizzle to cymbals or low-end thump to a bass guitar (respectively).

- Frequently **DEACTIVATE** the device to make sure that the changes you're making are actually improving the sound of the track and helping it to sit better in the overall mix.

Dynamics Processing

Like EQ, dynamics processing is an essential part of the audio production workflow. Unlike EQ, dynamics processing is not particularly well known outside of audio production. It is fair to say that working with dynamics processors is a bit more complex than using EQ to boost the bass or cut the treble on your car stereo. But the basics of dynamics processing are pretty straightforward, and if you use your ears to dial in just the right settings you'll be working like a pro in no time!

Types of Dynamics Processors

Dynamics processing refers to the manipulation of the volume (or amplitude) of an audio signal. Many instruments benefit from dynamics processing, including vocals, guitar, bass guitar, and drums. Depending on the situation, dynamics processors can be used to control a signal level that is varying dramatically, or to isolate just the loudest part of a signal while reducing or eliminating the quiet parts.

Some types of dynamics processors include:

- **Compressors**—The compressor is the most common type of dynamics processor. It is used to reduce the loudest part of an audio signal, resulting in a more predictable dynamic range. Once the loudest parts have been reigned in, it becomes possible to boost the entire signal without fear of clipping.

- **Limiters**—A limiter is essentially a compressor with a very high ratio setting (see the "Basic Dynamics Parameters" section below). As result, the incoming audio signal is completely "capped" at a specified level that prevents clipping. Limiters are an essential part of the music production workflow and are often used to make an overall mix seem louder.

- **De-Essers**—A de-esser is a compressor that operates at a specific frequency. Its primary use is to reduce excessive "s" sounds in a recording, a phenomenon known as *sibilance*. This type of processor can also be used to eliminate other undesirable high-frequency sounds such as the breath noise in a flute recording.

- **Expanders**—An expander is a dynamics processor that can be used to reduce the level of a signal when it falls below a certain level. Expanders work something like a compressor in reverse. A gentle amount of expansion can help to reduce some of the ambient noise that gets picked up in an audio recording.

- **Gates**—A gate (often called a "noise gate") is a type of expander that will completely silence a signal when it falls below a certain level. Gates are frequently used to eliminate the "bleed" that occurs when a microphone picks up signal from an "off-mike" sound source, such as a snare drum being picked up by a kick drum mike. Gates are also useful to eliminate background noise in a recording, such as the buzz of a guitar amp that can be heard in "silent" parts of the performance. The purpose of a gate is to clean up the sound of a recording by helping to isolate the desired parts of the audio signal and eliminate the unwanted parts.

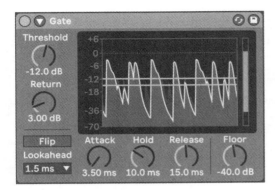

Figure 9.8 The Gate device in Ableton Live

Basic Dynamics Parameters

Just like EQ processors, dynamics processors make use of a standard set of controls. These will vary somewhat based on the type of dynamics processor being used.

Common dynamics parameters include the following:

- **Threshold**—The threshold parameter sets the level that the input signal must reach in order to trigger the processor. Note that compressors and limiters are triggered when the signal exceeds the threshold, whereas expanders and gates are triggered when the signal drops below the threshold.

- **Ratio**—The ratio parameter determines the amount of signal reduction that occurs once the threshold has been reached. For example, when using a compressor with the ratio set to 2:1, the signal must exceed the threshold by 2 dB to result in an output increase of 1 dB. A limiter typically features a compression ratio of 100:1.

- **Knee**—The knee parameter determines how suddenly or gradually gain reduction is introduced in a compressor as the signal level approaches the threshold.

- **Gain**—The gain parameter is used to add a level increase (known as "makeup gain") to a signal that has been reduced in volume by a dynamics processor.

- **Attack/Release**—The attack parameter determines how quickly the dynamics processor will act once the audio signal has reached the threshold. The release parameter determines how quickly the dynamics processor will disengage once the threshold is no longer met.

Dynamics Devices in Ableton Live

Ableton Live offers three types of dynamics plug-ins: compressors, limiters, and gates. The Compressor device is a transparent compressor that can also be used as a limiter by setting the ratio very high. The Limiter device features a ceiling control and a look-ahead buffer that can be combined to achieve a desired volume level without clipping. The Gate device can be used to gently reduce quiet sounds such as signal bleed or room noise, or to cut them out entirely.

Applying dynamics processing to tracks can seem confusing at first. But, just like when working with EQ, there are some common strategies to help you find the right settings.

Strategies for Using Compression

To add compression to a track in Ableton Live, you can add the Compressor device on the target track.

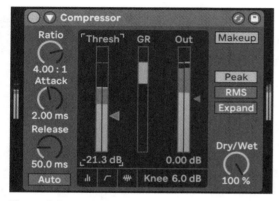

Figure 9.9 The Compressor device in Ableton Live (default view)

To use a Compressor on a track:

1. Set the **KNEE** to about 10 dB, the **ATTACK** to about 10 ms, and the **RELEASE** to about 100 ms.

2. Start off with a **RATIO** of about 3.0:1. (You may need a higher ratio if the track has a wide dynamic range.)

3. Reduce the **THRESHOLD** setting until you start seeing a small amount of compression registering in the gain reduction (GR) meter. (A good starting point is to target around 3 to 6 dB of gain reduction in areas of peak loudness).

> ⓘ You can use the small grey arrow to the right of the input meter to adjust the threshold while viewing the incoming signal level.

4. Adjust the **OUTPUT GAIN** control to increase the overall level of the track.

    Make sure that you don't increase the gain so much that the plug-in's output begins clipping. If you're having trouble finding the perfect gain setting, you may need to increase the ratio or lower the threshold.

Reverb and Delay Effects

Reverb and delay effects are indispensable tools for creating a sense of space. Collectively known as "time-based effects," reverb and delay can be used to create a wide range of sounds. This range spans the gamut from providing tasteful spatialization of audio, to adding a subtle spaciousness to a track, to creating dramatic effects that completely alter the sound of a voice or instrument.

What Is Reverb?

Reverb effects are used to create the sound of a real or imaginary space. Reverb algorithms can be used to recreate the sound of small spaces (rooms, cars, phone booths) or large spaces (concert halls, parking garages). A number of types of reverbs are available in both hardware and software formats. Some use a series of delays (see "What Is Delay?" below) to create smooth, musical-sounding results. Other reverbs use a technique called "convolution," which actually uses samples of real-world acoustic spaces to create incredible realism.

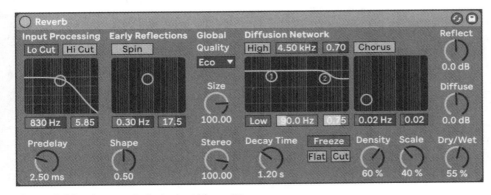

Figure 9.10 The Reverb device in Ableton Live

Reverb in Ableton Live

Ableton Live offers a basic reverb device called Reverb. Reverb offers a simplified set of controls that are an excellent introduction into the world of reverb.

Some of the Reverb controls include:

- **Quality**—The Quality control can be used to control the complexity of the reverb algorithm which determines the amount of CPU processing power used by the device.

- **Size**—The Size control is used to set a smaller- or larger-sounding space.

- **Decay Time**—The Decay Time control determines how long it will take for the reverb to "fade out" after a sound has been processed.

- **Lo Cut**—The Lo Cut control can be used to reduce the decay time for low frequencies.

- **Hi Cut**—The Hi Cut control can be used to reduce the decay time for high frequencies.

- **Predelay**—The Predelay control adjusts the amount of time that the reverb "waits" after receiving an input signal before it begins to create reverb.

- **Diffuse**—When set to a high value, the diffusion control will emphasize the initial build-up of echoes. Low values will reduce the initial build-up, which can result in better clarity.

- **Dry/Wet**—The Dry/Wet control can be used to adjust the dry/wet balance of the device's output. (See "Wet Versus Dry Signals" below.)

Applications for Reverb Processors

There are many creative ways to use reverb in a project. Whether adding a subtle sense of space to a vocal or completely washing out a guitar to create an ambient bed, reverb can truly bring a track to life.

Suggestions and scenarios for reverb usage:

- Add some character to a snare track by inserting a reverb directly on the track. Then adjust the wet/dry control to get the proper balance.

- Give a vocal track a vintage sound by using a medium reverb with the hi cut control enabled.

- Create a distant, washed-out guitar sound by using a large reverb and setting the wet/dry control to 100% wet.

- Add a medium reverb in a send-and-return configuration to apply reverb to a number of tracks in a session including vocals, acoustic guitars, keyboards, and more. This can help "gel" the mix by putting all of the components into the same acoustic space.

What Is Delay?

Delay is the term that audio engineers use for a processor that creates an echo. Delay is a simple effect in concept, but over the years both hardware and software delays have evolved into extremely complex processors.

Delay devices provide dozens of different approaches to delay, from digital models of vintage tape-style echos to modern granular pitch-shifted delays. And almost all modern delays have the ability to synchronize their individual echoes to the tempo of a song.

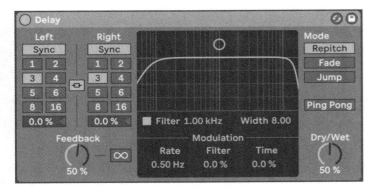

Figure 9.11 The Delay device in Ableton Live

Delay in Ableton Live

All editions of Ableton Live 10 feature a delay device called Delay. Delay is a simple but powerful delay plug-in that can generate anything from single echoes to beat-synchronized ping-pong delays that bounce back and forth between the left and right channels. The Delay device includes a pitch modulator option that can be applied to the delayed signal.

Controls in Delay include the following:

- **Delay Time**—The Left and Right delay time controls can be used to set delay iterations using note values that are synced to tempo or set in milliseconds.

- **Feedback**—The Feedback control determines how much of the plug-in's output will be "fed back" into the input. The feedback setting ultimately controls the number of echoes that will occur.

- **Filter**—The Filter controls can be used to set a range of frequencies to be delayed.

- **Modulation**—The Modulation parameters (Rate, Filter, and Time) can be used to control filter modulation.

- **Mode**—The Mode parameters (Repitch, Fade, and Jump) can be used to set the behavior of the device when the delay time is changed.

- **Ping Pong**—When enabled, the Ping Pong control alternates the delay repetitions between the left and right outputs of the device.

- **Dry/Wet**—The Dry/Wet control can be used to adjust the dry/wet balance of the device's output. (See "Wet Versus Dry Signals" below.)

Applications for Delay Processors

A little delay can help add excitement to your mix. This is a great way to propel a track forward and add rhythmic variety.

Suggestions and scenarios for using delay:

- Add some width to a mono acoustic guitar track by applying a delay with slightly different delay times for the left and right channels.

- Apply a simple quarter-note delay in a send-and-return configuration to add sustain, rhythmic emphasis, and interest to a variety of track types including vocals and guitar solos.

- Get an Edge-like guitar sound (U2) by running a simple guitar part into a dotted-eighth-note delay synced to the song's tempo. Set the feedback parameter high enough to provide at least 2 to 3 repetitions of each note.

Wet Versus Dry Signals

When working with time-based effects such as reverb and delay, you will need to balance the original signal with the processed signal. Audio producers refer to the original signal as the "dry" signal and the reverberated or delayed signal as the "wet" signal. Let's take a look at some common ways to adjust the wet/dry balance.

Using Time-Based Effects Directly on Tracks

Although time-based effects are typically applied using a send-and-return configuration (see below), at times you may use a reverb or delay as an insert directly on the source track. For example, you may want to add reverb to a snare track to give it a unique sound, or you may want to create a swirling guitar wash, where the signal is 100% wet. In these scenarios, you can place the desired plug-in directly on the target track.

When applying a time-based effect directly to a track, the only way to adjust the wet/dry balance is to use the plug-in's controls. The location and labeling of these controls can vary from one device or plug-in to another. The Reverb and Delay devices that come with Ableton Live use a knob control labeled **Dry/Wet**.

Figure 9.12 Reverb's wet/dry balance is adjusted using the Dry/Wet knob.

The Dry/Wet control can be adjusted to make the plug-in's output more dry or more wet. You can also type directly into the number field at the bottom of the control and manually enter a numerical value from 0% (completely dry) to 100% (completely wet).

 As a general rule, you'll want to set the device to a relatively dry setting to emulate "natural" reverb or delay characteristics. Start with a low setting, such as 20% and adjust from there as needed.

Using Send-and-Return Configurations

Configuring your time-based effects using a send and return is typically preferable, for a variety of reasons. First, you'll often want to apply the same reverb settings to many tracks in your project, so using a single device is much faster to set up and adjust.

Second, because time-based effects can use significant CPU power, using a send and return will save on CPU processing. You can create one instance of a reverb or delay on a Return track and then use as many sends to the track as necessary to apply the effect for multiple tracks.

And third, each source track's direct output will remain completely dry, while the send level can be used to adjust the amount of wet signal produced. This makes it quick and easy to modify the wet/dry balance for a whole bunch of tracks in your project without needing to change the settings of the device itself.

 The Dry/Wet control for time-based effects on a Return track should be set to 100% wet since the processed signal will be combined with the original (dry) signals from the source tracks.

Although advanced send-and-return configurations are beyond the scope of this book, you can configure a basic setup with a few simple steps. Use the techniques outlined below to get started with a reverb send.

 The same techniques can be used to create a delay send. Simply modify the name you apply to the Return track and the device you assign to the track.

To create a send-and-return for reverb:

1. Select **CREATE > RETURN TRACK** or press **OPTION+COMMAND+T** (Mac) or **CTRL+ALT+T** (Windows). Note the letter assigned to the Return (A through L).
2. Rename the new track to something meaningful, like **Reverb**.
3. Insert the Reverb device on the return track.
4. Adjust the Dry/Wet control as needed to set it to **100%** (completely wet).

To set the send levels:

- For each of the tracks you wish to send to the reverb, turn up the corresponding Send control (A through L).

Review/Discussion Questions

1. What is the difference between devices and plug-ins in Ableton Live? (See "Processor Basics" beginning on page 220.)

2. What are three ways that you show/hide the Detail View? What are three ways that you can show/hide the Device View? (See "Viewing Devices on Tracks" beginning on page 221.)

3. How can you insert a device on a track? How can you remove a device? (See "Inserting and Removing Devices" beginning on page 221.)

4. How can you move a device from one track to another track? How can you duplicate a device? (See "Moving and Duplicating Devices" beginning on page 222.)

5. What is the purpose of the Device Activator button in the device controls? What is the purpose of the Hot-Swap Presets button? (See "Common Device Controls" beginning on page 223.)

6. In what location in the Browser are presets saved when you press the Save Preset button? (See "Common Device Controls" beginning on page 223.)

7. Which keyboard modifier can you use to make fine adjustments on device controls? How can you quickly reset a device control to its default value? (See "Adjusting Device Parameters" beginning on page 225.)

8. In what ways can device parameters be adjusted using the computer keyboard? (See "Adjusting Device Parameters" beginning on page 225.)

9. What are the four common types of EQ bands? (See "Types of EQ" beginning on page 226.)

10. What three EQ device options are included with Ableton Live? (See "EQ Devices in Ableton Live" beginning on page 227.)

11. What three types of dynamics devices are found in Ableton Live? (See "Dynamics Devices in Ableton Live" beginning on page 231.)

12. What is the difference between reverb and delay? What category of effects do both of these processes belong to? (See "Reverb and Delay Effects" beginning on page 233.)

13. What are some scenarios in which you might place a time-based effect directly on a track? (See "Using Time-Based Effects Directly on Tracks" beginning on page 236.)

14. What are some advantages to using a send-and-return configuration for time-based effects? (See "Using Send-and-Return Configurations" beginning on page 237.)

 To review additional material from this chapter and prepare for certification, see the Ableton Live Audio Production Basics Study Guide module available through the Elements|ED online learning platform at ElementsED.com.

Exercise 9

Optimizing Tracks with Signal Processing

🎧 Activity

In this exercise, you will use devices to optimize tracks in Ableton Live. You'll begin by looking at some practical applications for equalization (EQ). Then, you'll explore some options for controlling dynamics. Finally, you'll create a send-and-return configuration to apply reverb for multiple tracks.

⏱ Duration

This exercise should take approximately 20 minutes to complete.

⊕ Goals/Targets

- Use the Channel EQ device to cut bass frequencies and boost high frequencies
- Use the Compressor device to reduce variations in peak volume levels
- Set up a send to a Return track
- Apply reverb processing using the Reverb device
- Save your work for use in Exercise 10

Exercise Media

This exercise uses media files taken from the song, "Lights," provided courtesy of Bay Area band Fotograf.

Written by: Zack Vieira and Eric Kuehnl; Performed by: Fotograf

The media provided for this course may be used for educational purposes only. No rights are granted to use the media for any other personal, commercial, or non-commercial purposes.

Getting Started

To get started, you will open your completed project from Exercise 8. This will serve as the starting point for this exercise.

Open your existing Lights project:

1. Launch Ableton Live, if it isn't already running. The Default Set will display.
2. Select **FILE > OPEN LIVE SET**.
3. Navigate to the location where you saved your work in Exercise 8 and select your **Lights03-xxx** project.
4. Click **OPEN**. The Set will open as it was when last saved.
5. Choose **FILE > SAVE LIVE SET AS** to save a copy of the Set under a new name.
6. In the resulting dialog box, name the copy **Lights04-xxx**, where xxx is your initials, and click **SAVE**.

Applying EQ and Dynamics to Tracks

To get an idea of where EQ processing might be useful, you should take a moment to listen to the project in its current state. Pay particular attention to any areas where frequencies are clashing between multiple tracks. Also listen for tracks that could benefit from more controlled dynamics.

Evaluate the tracks:

1. Press the **SPACEBAR** to begin playback.
2. Listen for tracks that compete for space in the same frequency range, such as the **Kick** track and the **Synth Bass** (or **Basses**) track. Try to formulate some ideas about how to give each track its own space in the frequency spectrum.
3. Also, listen for tracks with dynamics that might cause problems. Listen for consistency in the volume of notes for tracks that should serve as a steady backbone, such as when the bass guitar comes in during the choruses.
4. Feel free to solo and mute (deactivate) tracks to isolate certain elements. Feel free to adjust levels to get each track to cut through; see if you can identify certain tracks that present unique problems.
   - Is it hard to separate certain tracks from one another without causing one to obscure the other in a given frequency range?
   - Is it hard to find a level for certain tracks that allows all of the notes to be heard without causing certain notes to get too loud?

Adding EQ Processing

When using EQ, you may find yourself wanting to boost frequencies to emphasize the best parts of a track. However, it is equally important to cut frequencies to create room for other tracks to be heard. In this section of the exercise, you will use EQ to reduce the bass frequencies in the **Synth Bass** track (or **Basses** track) while slightly boosting the high frequencies in order to separate it from competing bass frequencies in the **Kick** track.

Insert an EQ plug-in on the Synth Bass track:

1. Locate the Channel EQ device in the Browser (Categories > Audio Effects > Channel EQ).

2. Drag and drop the Channel EQ device onto the **Synth Bass** track.

> If you are using the Set for Ableton Live Lite, drop the Channel EQ device on the **Basses** track instead.

Adjust the EQ settings:

1. Enable the high-pass filter band by clicking on the **HIGHPASS FILTER** button so that it becomes lit in yellow.

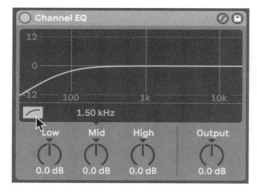

Figure 9.13 Clicking the Highpass Filter button (shown active)

2. Set the parameters for each EQ band:

   - **Low:** set the Gain to around -3.0 dB.

   - **Mid:** set the Frequency to 800.0 Hz (click and drag on the displayed value); set the Gain to around -2.0 dB.

   - **High:** set the Gain to around 2.0 dB.

   (See Figure 9.14.)

Optimizing Tracks with Signal Processing **243**

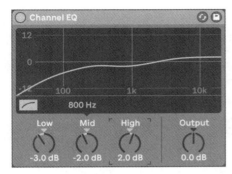

Figure 9.14 The Channel EQ device settings for the Synth Bass/Basses track

Adding Dynamics Processing

Next you'll apply dynamics processing to help control the signal levels. The bass guitar part has a slightly wider dynamic range than ideal, making it harder to position in the mix without constantly adjusting the volume fader. Here you will rein in the dynamic variation using a compressor.

> (i) A great way to learn to use devices such as EQ and Dynamics is to start from a relevant factory preset.

Use a factory preset to compress the bass guitar:

1. Locate the Compressor device in the Browser (Categories > Audio Effects > Compressor).

2. Open the Compressor folder to reveal its presets (click on the triangle).

3. Drag and drop the **OPTO-HARD COMPRESSOR.ADV** preset onto the **Bass Gtr** track.

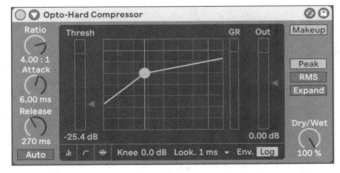

Figure 9.15 The Compressor device settings for the Bass Gtr track

> (i) If you are using Ableton Live Lite, place the Mildly Aggresive.adv preset onto the Basses track instead.

4. Switch to Arrangement View, if needed, and play through the first chorus of the song (starting around Bar 49). Listen to the bass guitar part. The dynamic variation should be dramatically reduced.

5. Select the **Bass Gtr** track (or **Basses** track); adjust the **Output Gain** using the triangle slider to the right of the Out meter on the Compressor device to find a good volume level for the track.

Using a Send-And-Return Configuration for Reverb

In this section, you'll add some reverb to help smooth out the strings and make them sound more natural. You'll set up a send-and-return configuration that can be used by multiple tracks in the project.

Create a new Return track:

1. Select **Create > Insert Return Track** or press **Option+Command+T** (Mac) or **Ctrl+Alt+T** (Windows). Note the letter assigned to the Return (A through L).

2. Select the track and press **Command+R** (Mac) or **Ctrl+R** (Windows); rename the track **Reverb**.

Assign a reverb device to the Return track:

1. Locate the Reverb device in the Browser (Categories > Audio Effects > Reverb).

2. Switch to the Session View, as needed; then drag and drop the Reverb device onto the **Reverb** track.

3. Click the **Hot-Swap Presets** button (circle with rotation arrows on the right side of the Title bar) to enter Hot-Swap mode. The button will change to show an **X** in red.

4. Select the **Reverb > Hall > Large Hall** preset in the Browser, and press **Return** (Mac) or **Enter** (Windows) to apply the preset.

5. Press the **Escape** key to exit Hot-Swap mode.

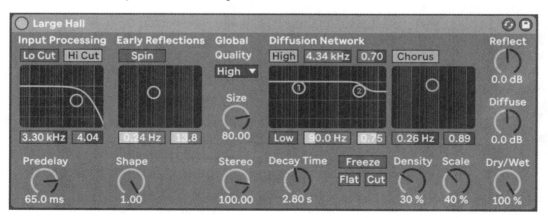

Figure 9.16 The Reverb device settings for the Reverb return track

Optimizing Tracks with Signal Processing **245**

Send audio from one or more source tracks to the Reverb track:

1. Locate the Send control corresponding to the Reverb return (A through L) on the **Strings** track.

> ⓘ **If you are using the Set for Ableton Live Lite, use the Send control on the Keyboards track instead.**

2. Increase the level of the send to around -3 dB. You should now be able to hear the reverberated strings when you play the second chorus of the project (starting around Bar 77).

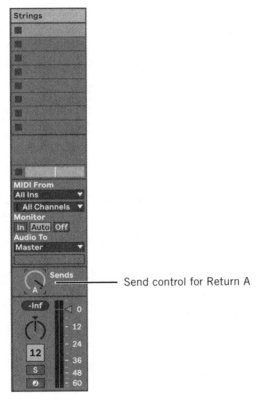

Figure 9.17 The Send control for the reverb send on the Strings track

3. (Optional) Use the Send control on the **Synth Pads**, **Synth Lead**, and **Piano** tracks to introduce some reverb for those instruments as well. Listen to the effect from the second chorus through the end of the piece and adjust the send levels to taste.

Finishing Up

To complete this exercise, you will need to save your work and close the project. You will be reusing this project in Exercise 10, so it's important to retain the work you've done.

Before you wrap up, you can also listen to the project to hear the mixing changes you've made.

Finish your work:

1. While the transport is stopped, click the **STOP** button to position the Insert Marker at the beginning of the arrangement.

2. (Optional) Press the **SPACEBAR** to begin playback and listen through the project. Press the **SPACEBAR** a second time when finished to stop playback.

3. Choose **FILE > SAVE LIVE SET** to save your work.

4. If desired, you can exit Ableton Live by doing one of the following:

    - On a Mac-based system, choose **LIVE > QUIT LIVE**.
    - On a Windows-based system, choose **FILE > EXIT**.

    Ableton Live will automatically close.

That completes this exercise.

Chapter 10

Finishing a Project

...What You Need to Do to Create a Stereo Mixdown...

This chapter introduces you to the processes of adding automation and creating a stereo mixdown of your Ableton Live project. We cover basic automation features in Ableton Live and how to use them. We also cover a variety of editing techniques that you can use to create or fine-tune the automation on your tracks. We then cover uses for limiters and dither plug-ins on the Master track, before wrapping up with a discussion on exporting your mix as a stereo file.

Learning Targets for This Chapter

- Understand what settings will and will not be recalled with a Ableton Live Set
- Learn how to enable automation recording
- Learn how to write automation changes
- Understand how to display and edit automation graphs
- Add appropriate processing to a Master track
- Use the Export Audio/Video command to create a stereo mix of a project

Key topics from this chapter are illustrated in the Ableton Live Audio Production Basics Study Guide module available through the Elements|ED online learning platform. Sign up for a free account at ElementsED.com.

After you have finished balancing your tracks, adding processing to the tracks, and adding any desired effects for the mix, it's time to start thinking about finishing up the project and creating a stereo mixdown. As you put the finishing touches on your mix, you can use automation to create and save dynamic changes. Then you can export your mix as a stereo file, configuring the file parameters as needed.

Recalling a Saved Mix

When you open an Ableton Live set that you've previously worked on, all of the mixer settings are recalled with the file, including the following:

- **Track layouts and views**—The track order, track sizes, and zoom settings that were in place when the project was last saved will all be recalled.

- **Mix attributes, settings, and automation**—The settings for Volume and Pan controls will be recalled for each track, along with any automation you previously created.

- **Track and send routing**—Any routing you've used for submixes, effects sends, and similar configurations will be recalled, along with the send levels.

- **Device assignments**—Device assignments and parameter settings will be recalled for each track.

Occasionally, however, conditions may exist that will prevent a project from recalling exactly as it was when last saved. These conditions can include the following:

- **External effects processors**—As discussed in Chapter 7, if you've used external gear for your mix, you will need to reconnect the gear and reconfigure all the hardware settings whenever you return to the project.

- **Unavailable Devices**—If you open the project from a system that does not have the same devices installed as the system where you created the project, any unavailable devices will be disabled.

- **Unavailable media**—Any audio or MIDI data that was not transferred will be missing when opening the project from a different system.

Automation

For a simple mix, you can often set the Volume and Pan controls as desired and leave them unchanged from the start of the mix to the end. For a more complex mix that requires dynamic changes during the course of playback, you can use Arrangement automation to change control settings throughout a song. Ableton Live allows you to record any real-time changes you make as automation on your tracks. Automatable controls include volume, pan, Track Activator (mute), send level, and more.

Arrangement Automation Workflow

In a typical production workflow, the ultimate destination for all automation is the Arrangement View. Aggregating all of the automation in a unified timeline enables a producer to fine tune automation while considering the needs of the entire song.

Automation can be recorded to the Arrangement View in two ways:

- Clip automation created in the Session View can be recorded to the Arrangement using the Arrangement Record function.
- Manual parameter changes can be recorded directly into the Arrangement.

Recording Arrangement Automation

Recording automation in the Arrangement View is quite simple. When the Automation Arm button is enabled, any track control changes made during an Arrangement Record pass will be stored as Arrangement Automation, regardless of the status of the individual track's Arm button.

To record Arrangement automation:

1. Enable the Automation Arm button in the Control Bar.

 — Automation Arm button

Figure 10.1 The Automation Arm button (yellow)

2. Click the Arrangement Record button (solid circle) in the Control Bar.
3. If needed, click the Play button to start the transport.

> ⓘ If the START PLAYBACK WITH RECORD preference is enabled, clicking the Arrangement Record button will immediately start playback.

4. Perform any desired control changes in real time during the record pass.

    If Automation Mode is currently enabled (see "Viewing Arrangement Automation" below), you will see the track's main automation lane update in real time. Otherwise, the track will turn red to show that automation is being written.

5. Stop playback when finished.

Automation LEDs

Any control that has automation data on it will display a red automation indicator (or "LED"). The location of this LED can vary depending on the parameter type and the current view. In the Session View, automation LEDs will appear to the upper left corner of most controls, and inside Volume sliders. In the Arrangement View, automation LEDs will appear in the upper left corner of the control.

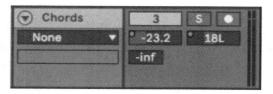

Figure 10.2 Pan and Volume Automation LEDs visible in the Session View (left) and the Arrangement View (right)

Viewing Arrangement Automation

By default, tracks in the Arrangement View show track content and not automation data. To view automation data, you must enable Automation Mode.

Toggling Automation Mode

To toggle Automation Mode on or off, do one of the following:

- Click the **AUTOMATION MODE** button.

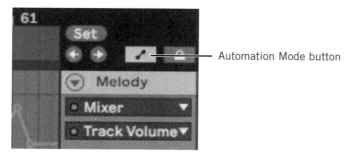

Figure 10.3 The Automation Mode button (blue)

- Choose **VIEW > AUTOMATION MODE**.

- Press the **A** key on the computer keyboard (with the Computer MIDI Keyboard disabled).

Viewing Automation Envelopes

With Automation Mode enabled, there are a number of ways to display automation envelopes.

An automation envelope is a horizontal line on the track showing a graphical representation of the changes for a particular parameter over time. Automation envelopes are also commonly referred to as automation graphs.

To view the automation envelope for a visible control:

- Click on any mixer or device control to display the control's envelope on the track. The automation envelope will appear in the track's main automation lane and will be superimposed on top of any audio or MIDI data.

To use the choosers to display an automation envelope:

1. Use the **DEVICE** chooser to select the track Mixer or a Device on the track.

2. Use the **AUTOMATION CONTROL** chooser to select the particular control that you would like to view. The automation envelope for the selected parameter will appear in the track's main automation lane and will be superimposed on top of any audio or MIDI data on the track.

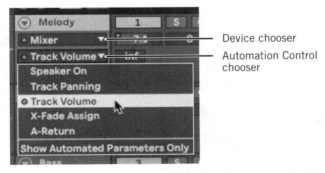

Figure 10.4 Selecting Track Volume from the Automation Control chooser

You can also hide all automation lanes on a track to prevent accidental changes to an automation graph while working in Automation Mode on other tracks.

To hide automation envelopes on the track's main automation lane:

- Click the Device chooser and select **NONE** from the top of the menu.

Drawing Automation Envelopes

In some scenarios, it may be necessary to manually draw an automation envelope in order to achieve a specific result. Ableton Live offers Draw Mode for these situations, which activates a pencil tool for drawing automation on screen.

To enable Draw Mode, do one of the following:

- Click the **Draw Mode** switch in the Control Bar.
- Press the **B** key on the computer keyboard.

Figure 10.5 The Draw Mode switch (yellow) with Draw Mode active

To momentarily toggle Draw Mode:

- Press and hold the **B** key on the computer keyboard; release to return to Edit Mode.

Drawing Automation

With Draw Mode enabled, you can simply click and drag in a track's Envelope Editor to create new automation data.

Drawing with Snap To Grid Enabled

Drawing with the Snap To Grid enabled will always result in "steps" that are the width of the current Edit Grid Value.

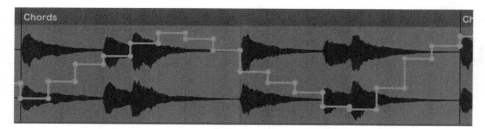

Figure 10.6 "Stepped" automation drawn with Snap to Grid enabled

To draw automation with a finer resolution:

- Hold **Shift** and then click and drag vertically on a segment.

Drawing with Snap To Grid Suspended

With Snap to Grid disabled, you can draw a "freehand" automation curve that is not affected by the current Edit Grid value. (See Figure 10.7.)

Figure 10.7 "Freehand" automation drawn with Snap to Grid disabled

To disable Snap to Grid, do one of the following:

- Press **COMMAND+4** (Mac) or **CTRL+4** (Windows).
- Right-click in the Arrangement View and set the Edit Grid Value to **OFF**.

To temporarily disable Snap to Grid:

- Hold **COMMAND** (Mac) or **ALT** (Windows) while drawing.

Editing Automation Breakpoints

With Draw Mode off, it becomes possible to use individual breakpoints to create and modify an automation envelope.

 Automation breakpoints are visible dots (nodes) along the automation graph where a parameter changes value or direction.

When creating or editing simple automation moves, it can be convenient to edit the breakpoints directly. As you move a breakpoint, a thin line will appear to show you the position of the breakpoint relative to the beat-time ruler, and a small box will appear to show you the vertical position of the breakpoint.

To add a new breakpoint, do one of the following:

- Click anywhere on a line segment. A new breakpoint will be added at the clicked location.
- Double-click anywhere in the envelope display. A new breakpoint will be added at the clicked location and the automation envelope will conform to include the new breakpoint location.

To delete an existing breakpoint:

- Click on an existing automation breakpoint. The breakpoint will be deleted and the automation envelope will adjust accordingly.

To move a breakpoint:

- Click and drag on an existing breakpoint. The breakpoint position will move and the automation envelope will adjust as you drag.

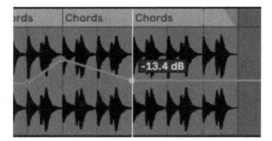

Figure 10.8 Moving a breakpoint

 Moving a breakpoint past a neighboring breakpoint will delete that breakpoint from the envelope. Moving the breakpoint back prior to releasing it will restore the neighboring automation breakpoint.

Deleting Automation

Sometimes you may simply want to delete all of the automation data for a particular control, device, or track. Delete operations are available for all of these situations.

To delete all automation data for a control, do one of the following:

- Right-click on the automated control and select **DELETE AUTOMATION** from the context menu.
- Right-click on an automation lane header and choose **CLEAR ENVELOPE**.
- Select an automated control and press **COMMAND+DELETE** (Mac) or **CTRL+DELETE** (Windows).

Once a control's automation data has been deleted, the automation LED will disappear and the control's value will return to a static value for the duration of the Arrangement timeline.

Figure 10.9 A control before deleting automation data (left) and after deleting automation (right)

Creating a Mixdown

Mixing down is the process of recording the output from multiple tracks to a stereo file. This process is also commonly referred to as *bouncing* the project. Mixing down is often the last phase of audio production, although you can create a bounce at any time to create a complete mix as a stereo file.

Adding Processing on the Master Track

As discussed in Chapter 9, a limiter is a dynamics processor that caps the audio signal at a specified level. A limiter can be used on the Master Track to protect against clipping. Using a limiter can also help make the overall mix seem louder. Here, we will use the Limiter device to function as a brickwall limiter.

 The Limiter device is not available in Ableton Live Lite. Use the Brick Wall device instead (Categories > Audio Effects > Compressor > Brick Wall).

To use the Limiter device on your mix in Ableton Live, do the following:

1. Locate the Limiter device in the Browser (**Categories > Audio Effects > Limiter**).

2. Drag and drop the device onto the Master track.

3. Play through the project, taking note of the readings displayed on the Limiter's **Gain Reduction** meter.

4. Set the Limiter's **Ceiling** to –0.10 dB. The limiter will not allow the signal exceed this level no matter how much gain is added.

5. Adjust the **Gain** control to add gain to the signal. Around 3.00 dB is a good starting point. This will increase the overall loudness of the mix while keeping the peak levels at or below the ceiling setting.

Figure 10.10 The Limiter device with the Gain and Ceiling controls set

In the Brick Wall device, you can generally keep the default settings unchanged. You may want to engage the Auto Release function (Auto button, bottom left) to have the device adjust the release time based on the incoming audio.

Considerations for Bouncing Audio

When exporting a mix from Ableton Live, the bounced file will capture all audible information in your mix just as you hear it during playback.

The following principles apply to the mixdown file:

- **The file will include only audible tracks.** What you hear during playback is exactly what will be included in the mix. Any tracks that are deactivated will not be included. Similarly, if any tracks are soloed, they will be the *only* tracks included.

- **The file will be a rendered version of your project.** Devices, sends, and external effects will be applied permanently. Listen closely to your entire project prior to exporting the mix to ensure that everything sounds as it should. Pay close attention to levels, being sure to avoid clipping.

- **The file length will default to your current arrangement selection.** If you have an active selection when you export your mix, the mix file will last for the length of the selection by default. If you do not have a selection, Ableton Live will create a bounce from the start of the project to the end of the longest track.

Exporting Audio

The Export Audio/Video command allows you to mix your entire Set directly to a stereo file on disk. The corresponding dialog box lets you set the bit depth, file format, and sample rate for the exported file. The export will automatically be rendered offline for faster-than-real-time mixdown.

Supported File Types

The file types that you can create from the Bounce to Disk command include the following:

- **WAV.** This is the default file format for Windows- and Mac-based audio production.

- **AIFF.** This file format is primarily used on Mac systems. Use the AIFF format if you plan to import the bounced audio into Mac applications that do not support the WAV format.

- **FLAC.** This is a lossless file format that can create smaller audio files without a perceptible reduction in sound quality.

- **MP3.** This file format can create smaller audio files for Internet streaming and portable devices. Use this file format to balance file size against audio quality.

Using the Export Audio/Video Command

The Export Audio/Video command combines the outputs of all currently audible tracks routed to the Master (or other selected option) to create a new audio file on your hard drive.

To export a mix of all currently audible tracks, do the following:

1. Adjust track output levels and finalize any automation. Any devices or effects settings that are active on your tracks will be permanently written to the exported media.

2. Make sure that all of the tracks you want to include in the mix are audible. To mix down all tracks in your Set, verify that no tracks are soloed or deactivated.

3. Verify that the output of each track is routed to the same destination (typically the Master track). Use the **OUTPUT TYPE CHOOSER** as needed to set the track outputs.

4. Choose **FILE > EXPORT AUDIO/VIDEO**. The Export Audio/Video dialog box will appear.

Figure 10.11 The Export Audio/Video dialog box

5. Verify that the **RENDERED TRACK** is set to Master (the default).

6. The **RENDER START** and **RENDER LENGTH** fields will already be populated by the current timeline selection (if one exists), or the start of the first clip in the arrangement until the end of the longest track (if no selection is present). You can also manually edit these fields to change the start and length of the exported mix.

7. Choose the desired sample rate for the exported file from the **SAMPLE RATE** pop-up menu. Higher sampling rates will provide better audio fidelity but will also increase the size of the resulting file(s).

> ⓘ If you plan to burn your mix directly to CD without further processing, choose 44.1 kHz as the sample rate for the bounce.

8. Enable the **ENCODE PCM** option and choose the desired file type for your exported file from the **FILE TYPE** pop-up menu. Available options include WAV, AIFF, and FLAC.

> ⓘ When Encode PCM is on, a lossless audio file is created. WAV, AIFF, and FLAC formats are available for PCM export.

> ⓘ You can also export to a WAV, AIFF, or FLAC file and simultaneously create an MP3 file. To do so, enable both the Encode PCM option and the Encode MP3 (CBR 320) option.

9. Choose the desired bit depth for the exported file from the **BIT DEPTH** pop-up menu.
   - Choose 16 if you plan to burn your mix to CD without further processing.
   - Choose 24 when you want to create a final mix that will be mastered separately.
   - Choose 32 for ultra-high-resolution files that will undergo further processing or editing.

> ⓘ The standard bit depth resolution for compact discs is 16 bits.

10. Select other options as desired.

11. After confirming your settings, click the **EXPORT** button. A Save dialog box will appear.

12. Specify a file name and save location for your exported file(s). By default, the exported file(s) will be named after the Ableton Live Set.

When performing an export, Ableton Live processes the export without audio playback. A progress window will appear, displaying the percentage of the export that has been completed. (See Figure 10.12.)

Figure 10.12 The Export Audio progress window

Creating a CD-Compatible Mixdown

Creating a CD-ready mixdown from an Ableton Live Set will require a bit-depth reduction during the bounce, to create a 16-bit file. To maintain the highest possible audio quality when reducing bit depth, it is necessary to apply *dither* to your audio.

Using Dither

Dither is a form of randomized noise used to minimize quantization artifacts in digital audio systems. Quantization artifacts are most audible in the lowest parts of a signal's dynamic range, such as during a quiet passage or fade-out.

Introducing dither to a signal helps to reduce quantization artifacts and preserve dynamic range by adding very low-level noise. Dither provides a trade-off between signal-to-noise performance and quantization artifacts. Proper use of dither can create better subjective performance out of 16 bits compared to audio without dither.

Adding Dither to the Mixdown

Dither processing can be added in Ableton Live during the export process when creating a mix at 16 or 24 bits. In these cases, the 32-bit floating-point output from Live will be reduced to create an audio file on disk at a lower bit depth. This in turn will reduce the dynamic range in the file on disk. Adding dither processing will help preserve some of the dynamic range that would otherwise be lost.

 Dither should be added only when performing a bit-depth reduction. DO NOT add dither to a mixdown if you are not reducing bit depth, as this will add unneeded noise to the mix.

To add dither during a mixdown:

1. Choose **FILE > EXPORT AUDIO/VIDEO**. The Export Audio/Video dialog box will appear.
2. In the PCM section of the dialog box, set the **BIT DEPTH** chooser to 16 or 24.

 For a CD-compatible mixdown, set the Bit Depth to 16.

3. Use the **Dither** chooser to select the desired dither option.

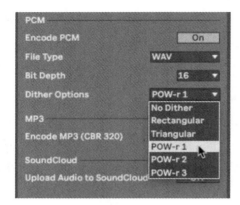

Figure 10.13 Dither options in the Export Audio/Video dialog box

4. Set other options in the dialog box as desired and click **Export** to complete the mixdown.

Types of Dither

When the dither setting is active in the Export Audio/Video dialog box, the **Triangular** option is selected by default. This mode is useful if there is a possibility that additional processing will be applied to the mix file, as it adds very little background noise. Rectangular mode introduces even less background noise, at the expense of additional quantization error.

The three **Pow-r** dither modes are generally the most effective at preserving dynamic range, offering successively higher amounts of dithering, but with the noise pushed above the audible range.

 If you will be sending your mixed file to a mastering house, you'll want to keep it at 32-bit floating-point resolution to retain the highest quality for the mastering process. In this case, you should NOT apply dither.

Review/Discussion Questions

1. What are some mix attributes that are recalled with a project in Ableton Live? What are some items that may not be recalled? (See "Recalling a Saved Mix" beginning on page 248.)

2. What control is used to enable real-time automation recording in the Arrangement View? (See "Recording Arrangement Automation" beginning on page 249.)

3. What are some ways that you can toggle Automation Mode on/off in the Arrangement View? (See "Viewing Arrangement Automation" beginning on page 250.)

4. What is the purpose of displaying an automation graph? What does an automation graph look like? (See "Viewing Arrangement Automation" beginning on page 250.)

5. Which Ableton Live feature activates a pencil tool that can be used to draw automation with the mouse? (See "Drawing Automation Envelopes" beginning on page 251.)

6. What are some ways of editing an automation graph? (See "Editing Automation Breakpoints" beginning on page 253.)

7. What is meant by the term *mixing down*? What other term is commonly used to describe this process? (See "Creating a Mixdown" beginning on page 255.)

8. What process can you use to optimize the output levels for your mix? What plug-in can you use in Ableton Live for this purpose? (See "Adding Processing on the Master Track" beginning on page 255.)

9. What are some things to consider when preparing to create a bounce of your mix? How will the bounce be affected by soloed or inactive tracks? (See "Considerations for Bouncing Audio" beginning on page 256.)

10. What sample rate and bit depth should you use when exporting a mix for use on an audio CD? (See "Creating a CD-Compatible Mixdown" beginning on page 259.)

11. What is the purpose of dither? When should dither be added to your mixdown? (See "Using Dither" beginning on page 259.)

 To review additional material from this chapter and prepare for certification, see the Ableton Live Audio Production Basics Study Guide module available through the Elements|ED online learning platform at ElementsED.com.

Exercise 10

Preparing the Final Mix

🎧 Activity

In this exercise, you will perform the steps necessary to create a final mix for your project in Ableton Live. You'll begin by adding processing to the Master track to optimize levels and prepare for exporting the mix. Then, you'll create a fadeout at the end of the song using real-time automation. Finally, you'll create a stereo mixdown of your finished project.

⏱ Duration

This exercise should take approximately 20 minutes to complete.

⊕ Goals/Targets

- Add a limiter to the Master track
- Work with Arrangement Automation
- Create a Stereo Mixdown

Exercise Media

This exercise uses media files taken from the song, "Lights," provided courtesy of Bay Area band Fotograf.

Written by: Zack Vieira and Eric Kuehnl; Performed by: Fotograf

The media provided for this course may be used for educational purposes only. No rights are granted to use the media for any other personal, commercial, or non-commercial purposes.

264 Exercise 10

Getting Started

To get started, you will open your completed project from Exercise 9. This will serve as the starting point for this exercise.

Open your existing Lights project:

1. Launch Ableton Live, if it isn't already running. After startup, the Default Set will be visible.
2. Select **FILE > OPEN LIVE SET**.
3. Navigate to the location where you saved your work in Exercise 9 and open your **Lights04-xxx** project. The Set will open as it was when last saved.
4. Choose **FILE > SAVE LIVE SET AS** to save a version of the project under a new name.
5. In the resulting dialog box, name the version **Lights05-xxx**, where xxx is your initials, and click the **SAVE** button.

Adding Processing to the Master Track

In this section of the exercise, you will add some processing to the Master track. By adding a limiter you will be able to precisely control the final mix levels without clipping.

 The Limiter device is not available in Ableton Live Lite. In this case, you can use the Brick Wall device instead.

Add a limiter to the Master track:

1. Locate the Limiter device in the Browser (Categories > Audio Effects > Limiter).

     In Ableton Live Lite, locate the Brick Wall device (Categories > Audio Effects > Compressor > Brick Wall).

2. Drag and drop the device onto the Master track.
3. Adjust the Limiter device settings as follows:
    - Gain: 4.5 dB
    - Ceiling: -0.20 dB

    (See Figure 10.14.)

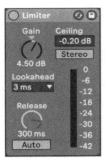

Figure 10.14 The Limiter device with the suggested settings

> In the Brick Wall device, set the Threshold around −2.5 dB, set Attack to 0.01 ms, and engage Auto Release (click the Auto button in the bottom left corner).

4. In the Arrangement View, go to Bar 73 and begin playback.

5. Watch the Gain Reduction meter in the Limiter device to see where the limiter is being activated. Allow playback to continue through the start of the chorus.

6. Switch to Session View and reset the **Master** track to its default level by double-clicking on the Volume control. You should now see more gain reduction occurring in the Limiter device, but no clip indication happening on the Master track's meters.

7. Continue monitoring playback through the end of the project to observe the results during the loudest sections and verify that the mix is not clipping.

8. Stop playback when finished.

Writing Automation

The music for this song has a sudden and unnatural cutoff at the end. You can address the problem by adding a final fadeout at the end of the song. Here, you'll record Arrangement Automation to write a fadeout in real time using the Volume control on the Master track.

Prepare to write automation on the Master track:

1. In the Arrangement View, click in a track near Bar 93 to position the Insert Marker.

2. Click on the **AUTOMATION ARM** button in the Control Bar, if not already enabled, so that it becomes active (highlighted).

Figure 10.15 Enabling Automation Arm in the Control Bar (shown active)

3. Click the Arrangement Record button (solid circle) to prepare to record. The button will turn red.

Automate a fadeout on the Master track:

> (i) To access the Volume control in Session View while keeping Arrangement View open, choose View > Second Window. Resize and position the second window to show the Master track controls without obscuring playback in Arrangement View.

1. Press the **SPACEBAR** to begin playback.

2. Click on the Volume control for the **Master** track and slowly move the fader all the way down to minus infinity (–∞). You'll have about 10 to 15 seconds to complete the fadeout.

3. Keep an eye on the Arrangement Position in the Control Bar (or in the Arrangement View) during your fadeout. Be sure to complete the fadeout before you reach Bar 101 where the audio ends. Once you have faded all the way down, you can release the Volume control and it will stay in position.

4. Stop playback once automation has been written a few bars past the end of the song.

5. Press the **SPACEBAR** to listen to the fadeout and verify the automation.

6. If you're not satisfied with the result, try another pass of writing the automation. It frequently takes several attempts to get the timing of a fadeout just right!

7. Once you're happy with the fadeout, disable the Automation Arm button.

View the volume automation on the Master track:

1. If needed, close the second window (or switch to Arrangement View) and locate the **Master** track.

2. Click the **AUTOMATION MODE BUTTON** above the tracks display to show track automation. The button will display in blue when enabled.

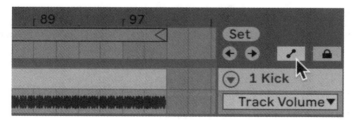

Figure 10.16 Clicking the Automation Mode button (shown enabled)

3. Click the **UNFOLD TRACK** button (triangle) on the Master track. The track will expand vertically, and the automation graph will show the results of the fadeout you wrote for the track. (See Figure 10.17.)

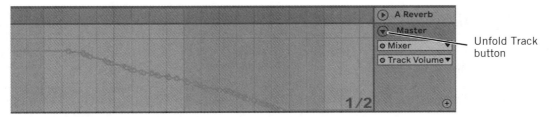

Figure 10.17 The volume automation graph on the Master track showing the final fadeout

4. Once you've confirmed that the automation graph looks correct on the Master track, click the **AUTOMATION MODE BUTTON** above the tracks display again to hide track automation.

Create a Stereo Mixdown

Use the following steps to export a stereo mixdown file for your project.

Export the mix to a stereo file:

1. Make a selection on any track, beginning at the start of the song and extending to slightly past the end of your fadeout (around Bar 102).

2. Choose **FILE > EXPORT AUDIO/VIDEO** to open the Export Audio dialog box.

3. Select the following options for your export:
   - Sample Rate: **44100**
   - File Type: **WAV**
   - Bit Depth: **16**
   - Dither Options: **POW-r2** (not available in Live Lite)

4. Leave the other options set to their defaults and click the **EXPORT** button. The Save dialog box will appear.

5. Name your file **YOURNAME-FINALMIX**.

6. Navigate to an appropriate save location, making note of the drive location and file path; then click **SAVE** to export your mix.

7. If needed, dismiss any new windows that open after completing the export.

Finishing Up

Congratulations! Over the course of these exercises, you have completed all of the required steps for an Ableton Live Set, including the following:

- Creating a set from scratch
- Importing and editing audio and MIDI files
- Mixing and balancing tracks
- Adding EQ, dynamics, and effects processing
- Recording automation
- Exporting the final mix to a stereo audio file

To wrap up, you'll need to locate the file you exported and verify the results.

Locate your exported stereo mix file:

1. Switch from Ableton Live to the Mac Finder or Window File Explorer.
2. Navigate to the folder location you selected for your export.
3. Select your mix file and do one of the following to listen to the results of your exported mix:
   - On a Mac system, press the **SPACEBAR** to begin playback using the QuickLook feature.
   - On a Windows system, right-click on the file and select **PLAY WITH WINDOWS MEDIA PLAYER**.
4. Listen through the entire mix to verify that the results are as you expected/intended.

 Be sure to listen to all of your mixed files as a quality check before sharing them with others.

5. If you hear any problem with the file (such as missing parts, wrong file duration, etc.), return to Ableton Live to correct the issue and repeat the export process.
6. Once satisfied, return to Ableton Live and choose **FILE > SAVE LIVE SET** to save your work.
7. Exit Ableton Live by doing one of the following:
   - On a Mac-based system, choose **LIVE > QUIT LIVE**.
   - On a Windows-based system, choose **FILE > EXIT**.

   Ableton Live will automatically close.

That completes this exercise.

Appendix A

Locating Missing Files in Ableton Live

At times, an Ableton Live Set file will open with one or more of its audio files (samples) missing. In this case, your tracks may show the message "Sample Offline" in the Arrangement View and/or in the Clip View. (See Figure A.1.) This can happen under any of the following conditions:

- The Set file has been separated from the associated Project folder: the Set file has been moved outside of its Project folder or transferred to another system using the Save As command.

- The Samples folder or any of its contents has been moved, renamed, or deleted.

- Audio files that you have previously imported into the Set from an external location have been moved, renamed, or deleted.

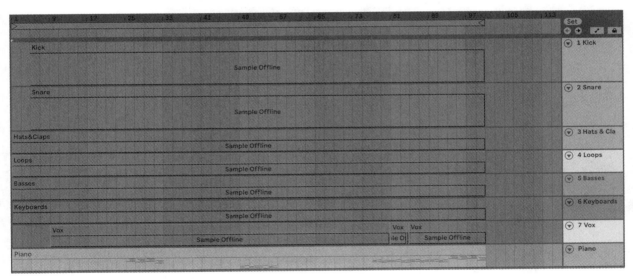

Figure A.1 Offline audio samples in the Arrangement View

On-Screen Information about Missing Files

When your Set file is not able to locate an audio file associated with the Set, a red warning message will appear at the bottom of the main window. (See Figure A.2.) Clicking on this warning message will open the File Manager in a panel on the right side of the main window. (See Figure A.3.)

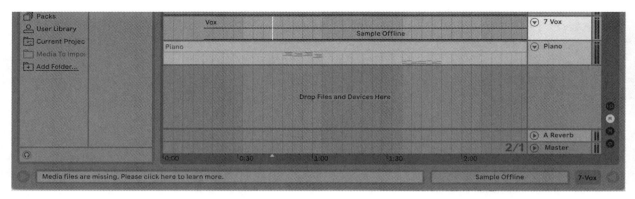

Figure A.2 Red warning message appears at the bottom of the main window; click here to open the File Manager.

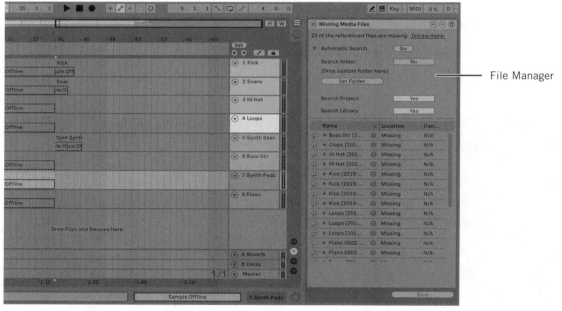

Figure A.3 The File Manager appears in the side panel on the right of the main window after clicking the warning message.

The File Manager can be used to replace missing audio files, provided that the files you need are available somewhere on your system.

The File Manager provides two methods of finding and replacing files. If you know specifically where each missing file is located on disk, you can perform a manual repair by drag-and-drop. If you do not know exactly where the missing files are, you can perform an automatic search for the files.

Replacing via Drag-and-Drop

To manually repair a Set with missing files, you can drag and drop individual replacement files from the Ableton Live Browser (or from a File Explorer or Mac Finder window) directly onto the associated file names listed in the File Manager.

This process can be tedious, especially if you have several missing files, as you have to locate the exact files to replace each of the missing files. If you drag a replacement file to the wrong item listed in the File Manager, you can end up with incorrect substitutions in the Set.

Replacing via Automatic Search

Arguably, a better solution is to use the Automatic Search function available at the top of the File Manager.

To use Automatic Search:

1. Click the triangle to the left of the **Automatic Search** heading, as needed, to unfold the section and reveal the search options.

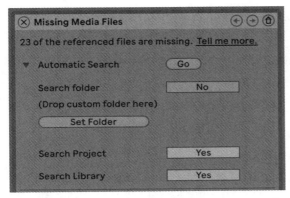

Figure A.4 Unfolding the Automatic Search section shows available search options in the File Manager.

2. If you know the general location where the missing files are on disk (such as within an older **Project** folder or inside a **Samples** folder in another location), you can use that location as a Search Folder:

    - Use the Ableton Live Browser or a File Explorer or Mac Finder window to navigate to the Project folder, Samples folder, or other location containing the missing files.

    - Drag the folder from the Browser (or File Explorer or Finder window) to the location labeled (**Drop custom folder here**) in the File Manager.

    If you do not know where the missing files are located, you can try running a search focused on either the current Project folder or the Live Library, or both. These locations are generally included by default. (Make sure **Search Project** and **Search Library** are both set to **YES**.)

If you have no idea where your missing files are, you can navigate to a top-level folder (such as your User folder or the root level of an attached external drive) and drag that folder to the drop location in the File Manager.

3. Click the **Go** button in the File Manager to begin the search. (Make sure **Search Folder** is set to **Yes**.)

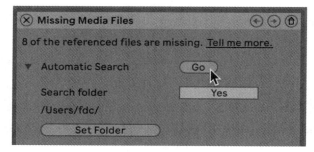

Figure A.5 Clicking rhe Go button will start a search in the designated locations.

A dialog box will display showing the search activity.

Figure A.6 Search activity dialog box

 Searching across an entire drive, User directory, or other high-level folder location can take a very long time, especially if the location contains a large number of files. It is generally better to target specific locations that are likely to contain the files you need, such as your Documents folder or, for the exercises in this book, the Live APB Media Files folder. This will produce much faster results.

Once the search completes, a dialog box will display showing the results.

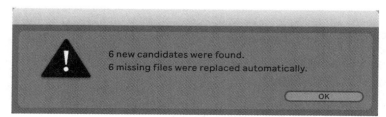

Figure A.7 Results dialog box after a successful search and replace operation

4. If some files are still missing after completing your search, you can either try a new search, choosing a different location, or continue working in the Set without the missing files.

Dealing with Multiple Candidates

In some cases, using the Automatic Search function will result in more than one potential match being found for a missing file in your Set. (See Figure A.8.) In this case, any missing files that have only a single match will automatically be replaced with the matching file. However, missing files that have more than one potential match, or candidate, will require you to choose between the candidates.

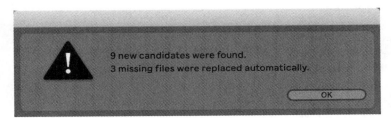

Figure A.8 Results dialog box after a search that resulted in multiple candidates for some files

Missing files that have multiple candidates available will display with a question mark icon (?) next to their file name in the File Manager. In the screenshot below, the Kick.wav, Loops.wav, and Snare.wav files each have two potential candidates to choose from.

Name	Location	Can...
▶ 06-Hats&Clap...	Project 06. ...	
▶ Basses.wav	Project 06. ...	
▶ Keyboards.wav	Project 06. ...	
▶ Kick.wav	Candidates	2
▶ Loops.wav	Candidates	2
▶ Snare.wav	Candidates	2
▶ Vox.wav	Missing	0
▶ Vox.wav	Missing	0

Figure A.9 Files shown in the File Manager preceded by a question mark have multiple candidates identified.

To select a candidate, do the following:

1. Click on the question mark icon next to the file you want to repair. The icon will change to show an X in red. This indicates that the file is in Hot-Swap mode. (See Figure A.10.) This will also cause the available candidate files to be displayed in the Browser. (See Figure A.11.)

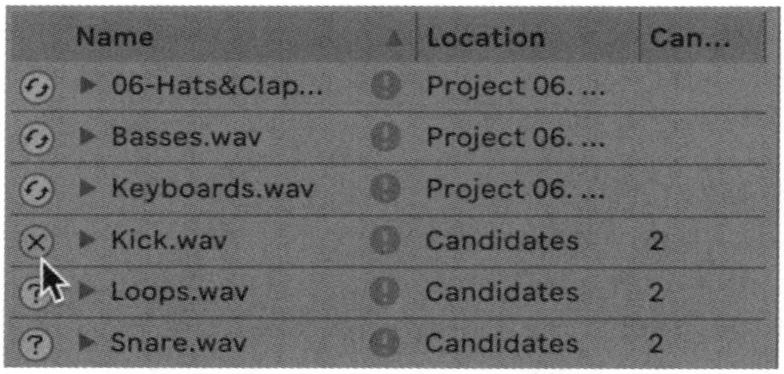

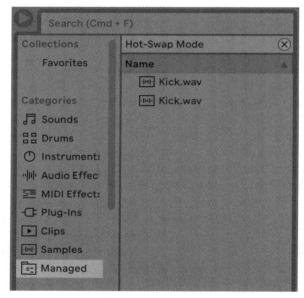

Figure A.10 The Kick.wav file in Hot-Swap mode

Figure A.11 The Browser will show the candidates that are available to replace the file you put in Hot-Swap mode.

2. Click on each of the candidates to audition them to identify the correct file.

   ⓘ You can also right-click on a file in the Browser to show it in the Finder / File Explorer or to see it in the Browser's Places view. Knowing the file location may help you determine whether it is the correct replacement file.

3. Once you've decided on the correct candidate, double-click on it in the Browser to use it as a replacement for the missing file.

   If your Set has additional missing files, you will be prompted with the option to replace other missing files with candidates from the same location. (See Figure A.12.) In many cases, a single location will contain most or all of the missing files.

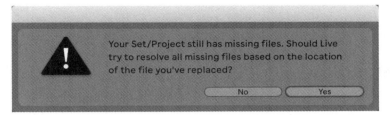

Figure A.12 Dialog box offering to search for other missing files in the same location

4. Select the desired option in the dialog box. If additional replacements are made, a second dialog box will display to indicate this.

5. Disable Hot-Swap mode for the file by clicking on the **X** icon in the File Manager. The icon will now display rotating arrows, indicating that the file is no longer missing.

Name	Location	Can...
▶ 06-Hats&Clap...	Project 06. ...	
▶ Basses.wav	Project 06. ...	
▶ Keyboards.wav	Project 06. ...	
▶ Kick.wav	Project 06. ...	
▶ Loops.wav	Candidates	2
▶ Snare.wav	Candidates	2

Figure A.13 Taking the Kick.wav file out of Hot-Swap mode

6. Continue selecting candidates for other missing files, as needed, using the procedures outlined in Steps 1 through 3 above.

Index

A

Ableton Live
  Ableton Live 10.app, 47
  AbletonLive10.exe, 47
  application icon, 47
  Launching, 47
Ableton Live Editions
  Feature Comparison, 40
Ableton Live Hardware
  Options, 41
Ableton Live Intro, xv, 7, 39–40, 44, 57, 77, 173
Ableton Live license, 47
Ableton Live Lite, 39, 44, 77, 85
Ableton Live Standard, xv, 39–41, 44, 57, 77, 173, 176, 187, 228
Ableton Live Suite, xv, 2, 8, 39–41, 44, 77, 173
Ableton Live Systems, 39
Activator, 140, 177, 248
  Using the Track Activator, 202
Aftertouch, 97
AIFF, 256, 258
Analog vs Digital Audio, 74
Arrangement
  automation, 249
  loop brace, 120, 123
  viewing automation, 250
Arrangement Loop, 122
Arrangement Overdub, 143
  button, 144
Arrangement Record, 142, 185, 249
  button, 141, 143, 185, 249
Arrangement View, 113
  loop, 122
  Loop switch, 123
  Scrolling, 125
  Zooming, 125

ASIO, 79
ASIO4ALL, 79
Audio
  batch importing, 178–179
Audio Basics, 66
  Amplitude, 67
  Frequency, 66
Audio Interfaces, 76
  ASIO, 79
  ASIO4ALL, 79
  Benefits, 10
  Budget, 78
  High-Resolution Audio, 77
  I/O Connectivity, 77
  Sound Quality, 77
  Working Without, 79
Audio Tracks, 120
Authorizing Live, 47–51
automation, 248
  arrangement, 249
  deleting, 254
  drawing, 252
  drawing envelopes, 251
  editing breakpoints, 253
  LEDs, 249, 254
  recording in arrangement, 249
  toggling Automation Mode, 250
  viewing envelopes, 250
  viewing in arrangement, 250
Automation Arm button, 249
Automation Breakpoints, 253–254
Automation Control chooser, 251
automation envelopes, 251
Automation Mode, 249–250
Auto-Monitoring, 141

B

Back to Arrangement button, 186
batch importing
    to new tracks, 179
    to tracks, 178
Beat-Time Ruler
    Zooming and Scrolling, 125
bit depth
    for compact discs, 258
bit-depth reduction, 259
Bouncing, 255
    Export Audio/Video command, 257, 259
Bouncing Audio
    Considerations, 256
Browser, 145
    Displaying the Browser, 145
    Manually Browsing, 146
    Searching for Files, 146
Buchla, Don, 88
Buffer Size setting, 53

C

Channel EQ, 227
clip automation, 249
Clip Launch button, 180
Clip Looping
    enabling/disabling, 182
clip slot
    record button, 183
Clip Slots Section, 173
Clip Stop button, 180
Clipping, 199
Clips
    Clips vs Files, 121
    Copying, 151
    Copying and Pasting, 149
    Cutting, 151
    Deleting, 149
    Moving, 149, 150
    Pasting, 151

    Selecting, 149
commands
    Export Audio/Video command, 256–257, 259
Compression
    Strategies, 232
Compressor, 230–232
    Using Compression, 232
Control Bar, 112
controls
    signal routing, 113
copy protection, 47
CPU, 5, 6, 233, 237
    Hyper-Threading, 6
    Processing Requirements, 7
    Processing Speed, 6
    Processor Cores, 6
CPU Load Meter, 51–52

D

DAW
    Ableton Live, 32
    Apple GarageBand, 34
    Apple Logic Pro X, 35
    Avid Pro Tools, 33
    Cakewalk Sonar, 37
    Cockos Reaper, 37
    Common DAWs, 32
    Functions, 32
    MOTU Digital Performer, 37
    Presonus Studio One, 37
    Propellerhead Reason, 37
    Steinberg Cubase, 36
    Steinberg Nuendo, 36
De-Esser, 230
Delay, 220, 233–237
    Applications, 235
    Controls, 235
    Definition, 234
    Delay Time, 235
    Dry/Wet, 235–236

Index

Feedback, 235–236
Filter, 235
Mode, 235
Modulation, 235
Ping Pong, 235
Usage, 236
Detail View
    Show/Hide, 221
Device Parameters, 225
    Adjusting with Keyboard, 226
    Adjusting with Mouse, 225
    Fine Adjustment, 226
    Return to Default, 226
Device View
    Show/Hide, 221
Devices
    Adjusting Parameters, 225
    Audio and MIDI Effects, 44
    Common Controls, 223
    Delay, 233
    Device Activator, 224
    Duplicating, 222
    Dynamics, 230, 231
    EQ, 226
    Hot-Swap Presets, 224
    Inserting, 221
    Moving, 222
    Removing, 221, 222
    Reverb, 233
    Save Preset, 225
    Viewing, 221
    Virtual Instruments, 44
Digital Audio
    Analog-to-Digital Conversion, 75
    Binary Data, 74
    Bit Depth, 76
    Dynamic Range, 76
    Sample Rate, 75
Dither, 247, 259–260
    adding dither, 259
Draw Mode, 157, 252–253

Duplicate Command, 152
Dynamics, 201, 230–232
    Attack/Release, 231, 232
    Basic Parameters, 231
    Compressors, 230
    De-Essers, 230
    Expanders, 230
    Gain, 231
    Gates, 230
    Knee, 231–232
    Limiters, 230
    Ratio, 231–232
    Threshold, 231–232
    Types, 230

E

Edit Commands
    Duplicate, 152
Edit Grid, 147–148
    Fixed Grid, 148
    Modes, 147
    Zoom-Adaptive Grid, 148
Editing
    Fading Clips, 154
    MIDI Clips, 153
    Non-Destructive, 152
    Resizing, 153
    Splitting a Clip, 153
Editing MIDI Clips, 155, 163
Editing Techniques
    Advanced, 152
    Basic, 149
Encode PCM, 258
EQ, 201, 226
    Basic Parameters, 227
    Filters, 226
    Frequency, 227
    Gain, 227
    Notch, 227
    Parametric, 226

Q, 227
Shelf, 227
Strategies, 229
Types, 226
Using, 229
EQ Eight, 227–229
EQ Three, 227–228
Expander, 230–231
Export Audio/Video
    command, 256–257
    dialog box, 257

F

File Hierarchy, 19
File Management, 11
    Creating Folders, 15
    Locating Files, 14
    Mac OS, 11
    Moving Files, 14–15
    Opening Files, 14–15
    Right-Click Functions, 15
    Using File Explorer, 13
    Using the Finder, 11
    Windows, 13
files
    AIFF files, 256
    MP3 files, 256
    WAV files, 256
Fixed Grid, 148
FLAC, 256, 258
Follow Mode, 124
Follow Scrolling Mode, 124

G

Gain-Based Processing, 205
Gate, 230–231
Group Tracks, 120

H

Height and Width
    Arrangement View, 126

I

I/O choosers, 175
Importing
    Audio and MIDI, 144
    Batch Importing, 178–179
    Drag and Drop Locations, 147
    from the Browser, 145
    From the Desktop, 145
    Supported Audio Files, 144
    To a New Track, 147
    To an Existing Track, 147
    to New Track, 178
    to Tracks, 178
Importing Clips, 178, 187
In/Out Section, 113
Input and Output
    In/Out Section, 175
Insert Devices, 221
Installation Requirements, 8
installers
    Ableton Live, 43
installing
    Ableton Live packs, 45–46

J–K

Keyboard Shortcuts, 21
    delete all automation data, 254
    narrow/widen the grid, 157
    scroll left/right, 160
    select all notes, 157
    Suspend Snap to Grid, 253
    toggle Automation Mode, 250
    toggle Draw Mode, 252
    Zooming and Scrolling, 126

L

Launching Applications, 17
    From the Desktop, 18
    Mac-Based Systems, 17
    Windows-Based Systems, 17
Limiter, 230–231, 255
    Usage, 255
Limiter device, 255
Loop
    Arrangement loop, 122
loop brace, 120, 123
Loop switch, 123

M

Mac Versus Windows, 3
    Mac-Based Computers, 3
    Running Windows on a Mac, 4
    Windows-Based Computers, 3
Main Views and Sections, 112
Master Fader
    Metering, 200
Master Track, 121
Metering, 199
    Master Fader, 200
Metronome, 138
Microphone Accessories
    Acoustic Shields, 71
    Connections, 72
    Mike Stand, 71
    Pop Filters, 71
    Shock Mounts, 71
    Vocal Booths, 71
    Windscreens, 71
Microphones, 67
    Accessories, 71
    Basic Setup, 72
    Basic Techniques, 70
    Condenser, 68
    Connections, 72
    Dynamic, 67

    Polar Pattern, 69
        Bidirectional, 70
        Cardiod, 70
        Omnidirectional, 70
    Position, 72
    Proximity Effect, 70
    Ribbon, 68
    USB, 68
MIDI, 88
    Cables and Jacks, 98
    Controller Messages, 91
    Creating MIDI Notes, 156
    Draw Mode, 157
    Editing, 155, 163
    Editing MIDI Notes, 158
    Editing Velocity, 159
    History, 88
    Interfaces, 99, 100
    MIDI Note Editor, 155
    MIDI versus Audio, 100
    Monitoring, 101
    Multiple Devices, 100
    Note Off, 90
    Note On, 90
    Notes, 90
    Plug-and-Play Setup, 98
    Program Changes, 91
    Protocol, 90
    Selecting MIDI Notes, 157
    Setup and Signal Flow, 98
    Velocity, 90
    What is MIDI?, 101
MIDI Arrangement Overdub, 143–144
    Enabling, 144
MIDI Clips
    Editing, 153
MIDI Controllers, 92
    Alternative Controllers, 96
    Digital Pianos, 92
    Drumpad Controllers, 95
    Grid Controllers, 95

Keyboard Controllers, 94
Purchasing, 97
Synthesizers, 92
What to Look for, 96
MIDI Note Editor
Navigating, 159
Scrolling, 160
Zooming, 160
MIDI Tracks, 120
Mixdown, 247–248, 255–256, 258–260
CD-compatible, 259
Creating a Mixdown, 255
Export Audio/Video command, 256
faster than real-time, 256
offline (faster than real-time), 258
Mixer Section, 113, 176
Mixing
Basic, 198
Clipping, 199
Clips vs Overs, 201
Deactivating Tracks, 202
In the Box, 207
Master Fader Metering, 200
Metering Source Tracks, 199
Panning, 202
Setting Levels, 198, 202
Using Solo, 202
Mixing in the Box, 207
Advantages, 207
Mod Wheel, 91, 97
Monitoring, 140
MIDI Controller, 142
Moog, Bob, 88
MP3, 256
Multi-Tracking, 72
Recording, 73
Signal Flow, 72, 74

N

Navigation, 121

Non-Destructive Editing, 152
Note Ruler
Scrolling, 160
Zooming, 160

O

Options
Mice and Trackballs, 10
Onboard Sound, 10
Trackpads and Touchpads, 10
Overview, 119, 124–126
Zooming and Scrolling, 125

P

Pan controls, 175
Panning, 202
Split Stereo Mode, 203
Stereo Pan Mode, 202
Pitch Bend, 97
Playback, 121, 122, 123
Arrangement Recording, 185
Controlling with Scenes, 181
Starting and Stopping, 180
Playhead
Tracking the Playhead, 124
Plug-Ins
AAX, 39
Audio Units, 39
Displaying Plug-In Windows, 223
Effects, 38
Formats, 38, 39
Virtual Instrument, 38
VST, 39
Preferences
audio device, 51
Auto-Hide Plug-In Windows, 223
Auto-Open Plug-In Windows, 223
Auto-Warp Long Samples, 165
Buffer Size, 53

default Live Set template, 120
optimizing performance, 52
Sample Rate, 53
scrolling behavior, 125
Select Next Scene on Launch, 181
Show Downloadable Packs, 46
Start Playback with Record, 141, 194, 249
Start Recording on Scene Launch, 185
Processing, 205
External Gear, 206
Gain-Based, 205
Master Track, 255
Time-Based Effects, 206, 233, 236
Track Devices vs. Sends, 206
Processors
Basics, 220
Punch-In Point, 139

Q

Quantization
recording quantization, 184
Quantization Menu, 184

R

RAM, 3, 4
Installing More, 5
Requirements, 5
Recalling a Saved Mix, 248
Recording
Auto-Monitoring, 141
from Session View to Arrangement, 185
in Session View, 183
Initiating a Take, 141
Metronome, 138
Monitoring, 140
on Scene Launch, 185
Quantization, 184
Setting Tempo, 138
Setting Up, 138

Signal Flow, 74
Stopping a Take, 141
to multiple tracks, 184
Undo and Delete Clip, 142
Using Arrangement Record, 185
Recording Audio, 139
Recording MIDI, 142
Monitoring a MIDI Controller, 142
Using MIDI Arrangement Overdub, 143
Resizing Views, 119
Return Tracks, 120
Reverb, 220, 233–237, 244
Algorithm, 233
Algorithms, 233
Applications, 234
Controls, 233
Convolution, 233
Decay Time, 233
Definition, 233
Diffuse, 234
Dry/Wet, 234, 236
Hi Cut, 234
Lo Cut, 233
Predelay, 234
Quality, 233
Size, 233
Usage, 234
Rulers, 119
Beat Time, 119

S

Sample Rate, 53
setting, 53
sample rates
for compact discs, 258
Saving Files, 18
Scenes Section, 173
Scrolling, 124–126, 129
Scrub Area, 119

Sections
    Clip Slots Section, 173
    In/Out Section, 175
    Mixer Section, 176
    Scenes Section, 173
    Send Section, 176
    Show/Hide, 118
    showing/hiding, 177
Selections, 122, 133, 149
Send controls, 176
Send selectors, 175
send-and-return, 219, 234, 236–237
Session Record button, 184
Session View, 172
    playback, 180
    Recording, 183
Show/Hide Sections, 118
Show/Hide Views, 118
Signal Flow, 72
Signals
    Wet vs Dry, 236
Solo
    Using Solos, 202
Storage, 7–9
    Cloud Storage, 9
    External Drives, 9
    Hard Disk Drive, 7
    HDD Versus SSD, 8
    Solid State Drive, 7

T

Tempo
    Setting Tempo, 138
    Setting Tempo with Scenes, 181
Time Signatures
    Setting Time Signatures with Scenes, 181
Time-Based Effects, 206, 233, 236
    As Inserts, 236
    Send-And-Return, 237
Track Activator, 140, 177, 202, 248

Tracks
    Adding, 120
    Audio Tracks, 120
    Group Tracks, 120
    Master Track, 121
    MIDI Tracks, 120
    Return Tracks, 120

U–V

Velocity Sensitivity, 97
views
    Arrangement View, 112–113
    Browser View, 112
    Clip View, 112
    Detail View, 112
    Device View, 112
    MIDI Note Editor, 155
    Session View, 112
    Show/Hide, 118
Virtual Instruments, 32, 35, 38–40, 55, 92, 97, 101–104, 120, 142–143, 221
    Assigning, 103
    Grand Piano, 168
    instrument devices, 38
    plug-ins, 6, 33–34, 38
    Tracking, 102
Volume controls, 175

W

WAV, 256, 258
waveforms, 113
Wet vs Dry Signals, 236
Working with an Application, 19
    Keyboard Shortcuts, 21
    Using Menus, 19

X, Y, Z

Zoom-Adaptive Grid, 148
Zooming, 125–126, 129, 133

About the Author

This book and associated coursework has been developed by **Eric Kuehnl** and **NextPoint Training, Inc.**, to prepare students for Ableton Live certification under the NextPoint Training Digital Media Production program.

Eric Kuehnl is a composer, sound designer, and educator. He is the Director of Partner Programs for NextPoint Training, and the Co-Director of the Music Technology Program at Foothill College. He also teaches game audio and audio post-production courses at Pyramind in San Francisco. Previously, Eric was an Audio Training Strategist in the Avid Education Department, and a Sound Designer for Sony Computer Entertainment America. He holds a Master's degree from California Institute of the Arts, a Bachelor's degree from Oberlin Conservatory, and studied composition at the Centre Iannis Xenakis in Paris.